Perfectly Matched Layer (PML) for Computational Electromagnetics

Perfectly Matched Layer (PML) for Computational Electromagnetics

Jean-Pierre Bérenger

ISBN: 978-3-031-00568-8   paperback
ISBN: 978-3-031-00568-8   paperback

ISBN: 978-3-031-01696-7   ebook
ISBN: 978-3-031-01696-7   ebook

DOI 10.1007/978-3-031-01696-7

A Publication in the Springer series

SYNTHESIS LECTURES ON COMPUTATIONAL ELECTROMAGNETICS #8

Lecture #8
Series Editor: Constantine A. Balanis, Arizona State University

Series ISSN: 1932-1252   print
Series ISSN: 1932-1716   electronic

First Edition
10 9 8 7 6 5 4 3 2 1

Perfectly Matched Layer (PML) for Computational Electromagnetics

Jean-Pierre Bérenger
Centre d'Analyse de Défense
Arcueil, France

SYNTHESIS LECTURES ON COMPUTATIONAL ELECTROMAGNETICS #8

ABSTRACT

This lecture presents the perfectly matched layer (PML) absorbing boundary condition (ABC) used to simulate free space when solving the Maxwell equations with such finite methods as the finite difference time domain (FDTD) method or the finite element method. The frequency domain and the time domain equations are derived for the different forms of PML media, namely the split PML, the CPML, the NPML, and the uniaxial PML, in the cases of PMLs matched to isotropic, anisotropic, and dispersive media. The implementation of the PML ABC in the FDTD method is presented in detail. Propagation and reflection of waves in the discretized FDTD space are derived and discussed, with a special emphasis on the problem of evanescent waves. The optimization of the PML ABC is addressed in two typical applications of the FDTD method: first, wave-structure interaction problems, and secondly, waveguide problems. Finally, a review of the literature on the application of the PML ABC to other numerical techniques of electromagnetics and to other partial differential equations of physics is provided. In addition, a software package for computing the actual reflection from a FDTD-PML is provided. It is available at www.morganclaypool.com/page/berenger.

KEYWORDS

Absorbing boundary conditions, Perfectly matched layer, Numerical method, Finite difference, Finite element, Free space, Stretched coordinate, Discretized space, Evanescent wave, FDTD, PML

Contents

Introduction

Nowadays, computers have been used for several decades to solve the partial differential equations of physics. To this end, numerous computational methods have been developed. In the field of electromagnetics, some, such as the asymptotic methods, solve an approximation of the Maxwell equations. Others solve the exact Maxwell equations numerically, or a set equivalent to the Maxwell equations. The latter methods are the most widely used. They can be grouped into two classes: firstly the methods based on the solution of integral equations, secondly the finite methods that solve the Maxwell equations in a direct manner by discretizing the physical space with elementary volumes.

The integral equations have been extensively used since the 1960's. They permit realistic problems of practical interest to be solved with relatively modest computers. The most known integral method is the method of moments developed by Harrington [1] in frequency domain. The integral equations are equivalent to the Maxwell equations, the boundary conditions, and the initial conditions of the problem to be solved. They are solved on part of the physical space reduced to a surface or a region of space, depending on the problem. These numerical techniques do not require absorbing boundary conditions (ABCs) and will no longer be mentioned in the following.

Several finite methods have been developed for solving the Maxwell equations in a discretized space. The most popular is the finite-difference time-domain method (FDTD) introduced by K. S. Yee [2]. The finite volume method (FVTD), the transmission line matrix (TLM) method, and the finite element method (FEM) are finite methods as well. With all these numerical techniques the physical space is split into elementary cells, elements, or volumes, that must be smaller than both the shortest wavelength of interest and the smallest details of the geometry of the objects to be placed within the part of space of interest. Since the computers are not able, and will never be able, to handle an infinite number of elementary cells or elements, these methods only allow the Maxwell equations to be solved within a finite part of space. This is inconsistent with the requirements of most problems of electromagnetics that are unbounded problems. Consider for instance two typical problems of numerical electromagnetics, first the calculation of the radiation pattern of an antenna, second the interaction of an incident wave with a scattering structure. In both cases the radiated field propagates toward the free space surrounding the structure of interest; in other words the physical boundary conditions should be placed at infinity. If the Maxwell equations are solved within a finite volume bounded with arbitrary conditions, the solution is erroneous. In order to overcome such contradictory requirements, that is a physical unbounded space to be replaced with a finite computational domain, the so-called absorbing boundary conditions have been introduced.

The absorbing boundary conditions (ABCs) simulate or replace the infinite space that surrounds a finite computational domain. The replacement is never perfect. The solution computed within an ABC is only an estimate to the solution that would be computed within a really infinite domain. Moreover, the ABCs cannot replace sources of electromagnetic fields, they only absorb fields produced by sources located inside the surrounded domain. From this, sources cannot be placed outside the ABCs. As a corollary, the ABCs can be implemented only upon concave surfaces.

Various ABCs have been developed over the years in the field of electromagnetics, from the extrapolation [3] or the radiating boundary [4] in the 1970's to the perfectly matched layer (PML) [5] and the complementary operators method (COM) [6] in the 1990's. This lecture is devoted to the presentation of the PML ABC, initially introduced in [5] for use with the FDTD method. Since then, the PML ABC has been the subject of numerous works reported in the literature, with the objective of improving it, extending it to other numerical techniques of electromagnetics, and extending it to the solution of partial differential equations governing other domains of physics, such as acoustics, seismic, or hydrodynamics. The lecture is organized as follows:

- Chapter 1 discusses the requirements that must be fulfilled by the ABCs in view of replacing a theoretical infinite space with a finite computational domain. This chapter also reviews the ABCs that were used before the introduction of the PML ABC.

- Chapter 2 introduces the PML concept in the two-dimensional case.

- Chapter 3 extends the PML ABC to three dimensions and to general media. The PML medium is interpreted in terms of stretched coordinates and dependent currents, and the complex frequency shifted stretching factor is introduced.

- Chapter 4 derives the different forms of time domain equations, namely the split PML, the CPML, the NPML, the uniaxial PML, for a vacuum, lossy media, and more general anisotropic and dispersive media.

- Chapter 5 is devoted to the FDTD method. The FDTD equations are provided for the various forms of PML media. Propagation and reflection of waves in the discretized FDTD-PML space are derived theoretically and discussed, with a special emphasize on the case of evanescent waves.

- Chapter 6 presents the application of the PML ABC to two typical problems of numerical electromagnetics solved with the FDTD method, namely a wave-structure interaction problem and a waveguide problem. The origin of spurious reflections from the PML is discussed and remedies are given so as to optimize the PML performance.

- Chapter 7 briefly reviews the extensions of the PML concept to other systems of coordinates, other numerical techniques, and other partial differential equations of physics.

CHAPTER 1

The Requirements for the Simulation of Free Space and a Review of Existing Absorbing Boundary Conditions

Answering two questions is the principal objective of this introductory chapter. The first question is: why is the simulation of free space needed in numerical electromagnetics? The second one is: which requirements have to be satisfied by the methods that simulate free space? In addition, the methods developed for simulating free space before the introduction of the perfectly matched layer concept are briefly reviewed.

1.1 THE MAXWELL EQUATIONS AND THE BOUNDARY CONDITIONS

The electric and magnetic fields E and H in material media are governed by the Maxwell equations

$$\nabla \times \vec{E} = -\mu \frac{\partial \vec{H}}{\partial t} \qquad (1.1a)$$

$$\nabla \times \vec{H} = \varepsilon \frac{\partial \vec{E}}{\partial t} + \vec{J} \qquad (1.1b)$$

with two Gauss laws satisfied at any time:

$$\nabla \cdot \mu \vec{H} = 0 \qquad (1.2a)$$

$$\nabla \cdot \varepsilon \vec{E} = \rho. \qquad (1.2b)$$

Permittivity ε and permeability μ are scalar quantities in isotropic media and tensor quantities in anisotropic media, J is a current density, and ρ is a charge density.

The Maxwell equations (1.1) are a set of two first-order partial differential equations connecting the time derivatives of E and H fields to some partial space derivatives of their

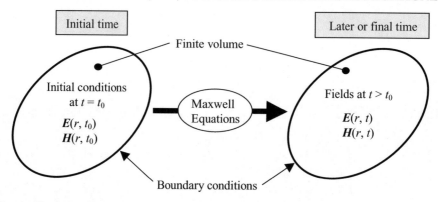

FIGURE 1.1: Evolution on time of the electromagnetic field governed by the Maxwell equations within a space domain bounded with boundary conditions

components. As known, this set of two equations can be merged into one second-order partial differential equation, namely the wave equation. As any partial differential equation or set of partial differential equations, the Maxwell equations are satisfied by an infinite number of solutions. In other words, there are an infinite number of physical problems that satisfy Eqs. (1.1). But there is only one solution that satisfies the following two additional conditions:

(1) *initial conditions*, that is E and H fields impressed within a given volume at an initial time,

(2) *boundary conditions*, that is E and H fields impressed at any time upon the whole surface enclosing the given volume.

The evolution in time of the initial E and H fields is governed by Eqs. (1.1) in conjunction with the boundary conditions. Initial E and H fields are physical fields that satisfy (1.2). It is trivial to prove, by multiplying (1.1) with nabla operator, that the evolution in time preserves the satisfaction of (1.2). Solving a problem of electromagnetics, especially by means of numerical methods, consists of using the Maxwell equations (1.1) to advance in time the electromagnetic fields within a given part of space bounded with impressed boundary conditions, from an initial time to a later final time. This is summarized in Fig. 1.1. In principle, the finite methods are well suited to the solution of such problems. The volume of interest is discretized with a finite number of elementary volumes, called cells or elements, depending on the method. Nevertheless, an important difficulty arises as using finite methods, because in most applications the domain is, at least in theory, of infinite extent. This is discussed in the following.

1.2   THE ACTUAL PROBLEMS TO BE SOLVED WITH NUMERICAL METHODS

Ideally, a problem well suited to finite methods is like in Fig. 1.1, with a domain of resolution of the Maxwell equations as small as possible, limited to the region of interest, that is the region where the field has to be computed. This allows the number of elementary volumes and then the number of unknown fields to be as small as possible, or alternatively the discretization of space to be as fine as possible. Unfortunately, most problems encountered in numerical electromagnetics significantly differ from this ideal case. In most cases the domain of interest is not bounded with an impressed boundary condition. Instead, the region of interest is open, at least in part, to the surrounding free space. This means that the boundary condition is rejected to infinity, or equivalently that the computational domain is, in theory, infinite.

A popular problem involving an infinite domain is the calculation of the radiation pattern of an antenna. Only fields in the vicinity of the antenna are needed—the far fields can be obtained by a near-field to far-field transformation—but the antenna radiates in the surrounding free space. If an arbitrary boundary condition is placed at a finite distance from the antenna, the radiated field is reflected toward the inner domain, resulting in the addition of a spurious field to the solution in the vicinity of the antenna. In theory, this difficulty could be overcome with time domain methods, by working with a large domain in such a way that the fields reflected from the arbitrary boundary enter the region of interest after the end of the calculation. In actual applications, such a solution is not realistic, because the required computational domain would be so large that the problem could not be handled by the computers. From this, for the calculation of the field near an antenna with a finite method, the infinite space surrounding the antenna must be replaced with an appropriate boundary condition placed at a distance as short as possible from the antenna. This boundary condition must allow the fields computed in the domain to be a satisfactory approximation to the fields that would be obtained if the computational domain were infinite. Such a boundary condition is called an absorbing boundary condition (ABC) because it must remove the reflection of fields toward the inner domain, that is the ABC must absorb the radiated outgoing fields.

Problems that are close to antenna problems are the calculations of the interaction of an incident wave with a structure of interest. Such problems include radar cross-section (RCS) calculations and electromagnetic compatibility (EMC) calculations. The field scattered by the structure is radiated toward the surrounding infinite space. An ABC placed as close as possible to the structure is needed so as to replace the infinite free space and allow the overall domain to be as small as possible. This permits the computational resources to be devoted to the use of a discretization of the structure as fine as possible.

Besides problems open in totality to free space, there exist some problems that are only partially open. Examples can be found in the field of waveguides where most of the computational domain is bounded with the walls of waveguides. The domain is in general only open in one direction, for instance at one end of the waveguide. Nevertheless, an ABC is also needed in such partially open problems so as to limit to a reasonable size the computational domain.

1.3   THE REQUIREMENTS TO BE SATISFIED BY THE ABSORBING BOUNDARY CONDITIONS

Let us consider the field radiated from a small dipole antenna. In spherical coordinates (r, θ, φ), the **E** and **H** fields are given by:

$$\overrightarrow{E}(r, \theta, \varphi) = \frac{-j I l e^{-j\omega r/c}}{4\pi \varepsilon_0 \omega} \left[2\cos\theta \left(\frac{1}{r^3} + \frac{j\omega}{c r^2} \right) \overrightarrow{u}_r + \sin\theta \left(\frac{1}{r^3} + \frac{j\omega}{c r^2} - \frac{\omega^2}{c^2 r} \right) \overrightarrow{u}_\theta \right]$$

(1.3a)

$$\overrightarrow{H}(r, \theta, \varphi) = \frac{I l e^{-j\omega r/c}}{4\pi} \sin\theta \left(\frac{1}{r^2} + \frac{j\omega}{c r} \right) \overrightarrow{u}_\varphi$$

(1.3b)

where ω is the angular frequency, l is the dipole length, and I is the magnitude of the current upon the dipole. As known, far from the dipole ($r \gg$ wavelength), the radiated field (1.3) is like a homogeneous plane wave whose magnitude decreases as $1/r$. Conversely, at distances of the order of, or shorter than, the wavelength, the field is not homogeneous and its magnitude rapidly decreases with distance.

The behavior of the field radiated by a dipole is general. Far from any radiating or scattering structure the field is like a plane wave in a vacuum, with a magnitude decreasing as $1/r$. This is known as the Sommerfield radiation condition. Conversely, in the vicinity of the structure the field is not homogeneous and rapidly decreases with distance and its form is complex. Especially, this is the case around scattering structures stricken by an incident pulse. Strongly evanescent fields are present at frequencies lower than the resonance of the structure, up to a distance of the order of its size.

Other problems where evanescent fields are present near the domain of interest, are waveguide problems. Within a waveguide, both traveling and evanescent waves can exist. Below a cutoff angular frequency ω_{cutoff} the TE and TM modes are evanescent in the longitudinal direction of the waveguide. As an example, within a parallel-plate guide each mode is the superposition of two waves whose space dependence is of the form:

$$e^{-\eta j \frac{\omega}{c} \cosh \chi y} e^{\eta \frac{\omega}{c} \sinh \chi x}$$

(1.4a)

where x and y are the longitudinal and transverse directions, $\eta = \pm 1$, and:

$$\sinh \chi = \pm \sqrt{\frac{\omega^2_{\text{cutoff}}}{\omega^2} - 1}, \qquad\qquad (1.4b)$$

with, for mode n and a guide of transverse size a:

$$\omega_{\text{cutoff}} = \frac{n\pi c}{a}. \qquad\qquad (1.4c)$$

From this brief overview of the fields radiated or scattered in typical open problems of numerical electromagnetics, it appears that the requirements that an absorbing boundary condition must satisfy strongly depend on its location with respect to the source of the field:

- if the ABC is placed far from the source, the ABC only has to absorb homogeneous plane waves propagating with the speed of light c. In general the plane waves strike the boundary at oblique incidence.
- if the ABC is placed in the vicinity of the source, the ABC must be able to absorb nonhomogeneous evanescent waves. One might think that this requirement is more severe than only absorbing homogeneous traveling waves.

Equivalently, the above can be reformulated as follows:

- if the ABC is only able to absorb homogeneous plane waves, it must be placed out of the evanescent region surrounding the source (antenna, scattering structure, waveguide).
- if the ABC is able to absorb evanescent fields, it can be placed close to the source, in the evanescent region. In that case, the overall computational domain is significantly smaller.

1.4   THE EXISTING ABCs BEFORE THE INTRODUCTION OF THE PML ABC

From a general point of view, there exist two categories of absorbing boundary conditions:

- the global ABCs based on the fact that the field at any point on the boundary of a given volume can be expressed as a retarded-time integral of the field upon a surface enclosing all the sources [7]. Such global ABCs are computationally expensive and are only marginally used in numerical electromagnetics [8].
- the local ABCs with which the field on the boundary is expressed as a function of the field in the vicinity of the considered point, that is in function of the field at the closest points of the mesh with finite methods. All the ABCs used in the past in computer

codes are local ABCs, and the perfectly matched layer ABC also can be regarded as a local ABC.

Various local ABCs have been proposed over the years, in parallel to the growing use of finite methods in numerical electromagnetics. Most conditions are designed for the absorption of traveling waves. Implicitly this means that they must be placed some distance from the sources, outside the evanescent region.

The ABC described in [4] for the FDTD method, denoted as radiating boundary, assumes that the field around a scattering structure is like the field radiated from a short dipole antenna. The field outside the computational domain, needed for the advance of the field on the boundary of the domain, is obtained from formulas of the dipole, assuming that the dipole is located at the center of the domain.

The ABC of Engquist–Majda [9], presented in 1977, is based on an approximation of the wave equation, valid for traveling waves propagating toward the boundary. The approximation is called a one-way wave equation. Different orders of approximation are possible, the first two ones read:

$$\frac{\partial E_u}{\partial x} - \frac{1}{c}\frac{\partial E_u}{\partial t} = 0 \tag{1.5a}$$

$$\frac{\partial^2 E_u}{\partial x \partial t} - \frac{1}{c}\frac{\partial^2 E_u}{\partial t^2} + \frac{c}{2}\frac{\partial^2 E_u}{\partial y^2} = 0 \tag{1.5b}$$

where E_u is any component of the $\boldsymbol{E}$ field. Equations (1.5) give the space derivatives of the field in the direction x normal to the boundary. With time-marching methods this permits the field on the boundary to be advanced in time. Notice that (1.5a) is satisfied rigorously by a plane wave striking the boundary at normal incidence. The discretized counterpart of (1.5a) in the FDTD method is nothing but an extrapolation in space and time, assuming that the field on the boundary is a plane wave at normal incidence. Equation (1.5b) takes account of the transverse derivative of the field, resulting in a better approximation for waves at oblique incidence. The reflection coefficient of ABCs based on one-way wave equations, for a plane wave at incidence θ, is:

$$r = \left(\frac{1 - \cos\theta}{1 + \cos\theta}\right)^n \tag{1.6}$$

where n is the order of the approximation, $n = 1$ with (1.5a) and $n = 2$ with (1.5b). The implementation of (1.5b) in the FDTD method was presented by Mur in 1981 [10]. This ABC, known as the Mur ABC, has been the most popular ABC in numerical electromagnetics for more than one decade. The Engquist–Majda ABC was generalized by Trefethen and

Halpern [11], so as to permit the nulls of the reflection coefficient to be placed at oblique incidences. This allows a small reflection to be achieved up to wide angles of incidence.

Another ABC later introduced [12] is the Higdon ABC based on a linear operator that differs from (1.5a) with the introduction of a term $\cos\alpha_i$. This allows the reflection of a plane wave to vanish at incidence α_i in place of normal incidence with (1.5a). Applying several times the operator, the n-order Higdon operator is obtained:

$$\left[\prod_{i=1}^{n}\left(\cos\alpha_i\frac{\partial}{\partial t}-c\frac{\partial}{\partial x}\right)\right]E_u=0 \tag{1.7}$$

whose reflection coefficient vanishes at the α_i angles. As with Trefethen–Halpern ABC, a small reflection of traveling waves can be achieved up to wide incidences.

Finally, the above ABCs allow excellent reflection coefficients to be achieved, especially by using a high order ABC with $n > 2$. Nevertheless, the Achille tendon of all these ABCs is in the fact that they were designed to absorb homogeneous traveling waves. This is not well suited to the realistic problems to be solved by means of numerical techniques, because in most physical problems the electromagnetic field in the region of interest involves both traveling and evanescent waves. This explains why the global performance of the one-way wave and operator ABCs cannot be improved significantly by increasing the order of the ABC over order 2. Especially, these ABCs cannot be placed close to the sources, that is in the evanescent region, whatever may be their order. In most cases this results in a computational domain significantly larger than the region of interest, i.e., the domain is larger than the optimum domain that would be used with an ideal ABC that could absorb both traveling and evanescent waves and could be placed nearby the region of interest.

Another ABC based on a quite different principle has been used with the FDTD method [13, 14]. This ABC is based on a matched medium that absorbs without reflection the electromagnetic waves striking the vacuum–medium interface at normal incidence. A layer of this medium is placed in the outer FDTD cells of the computational domain, so as to absorb the outgoing waves. In this matched medium, the Maxwell equations are replaced with:

$$\varepsilon_0\frac{\partial\overrightarrow{E}}{\partial t}+\sigma\overrightarrow{E}=\nabla\times\overrightarrow{H} \tag{1.8a}$$

$$\mu_0\frac{\partial\overrightarrow{H}}{\partial t}+\sigma^*\overrightarrow{H}=-\nabla\times\overrightarrow{E} \tag{1.8b}$$

where σ^* is a nonphysical parameter that allows the absorption of the magnetic field to be symmetrized with respect to the absorption of the electric field, provided that the following relationship holds:

$$\frac{\sigma}{\varepsilon_0} = \frac{\sigma^*}{\mu_0}. \tag{1.9}$$

With condition (1.9), called the matching condition, the impedance of a plane wave in the medium equals the impedance in a vacuum. This results in zero reflection at an interface between a vacuum and this medium, at normal incidence only. More precisely, in the conditions where the conductivity is large enough so as to realize an actual ABC [14], the reflection coefficient is identical to the first order of (1.6). This matched layer ABC has been used mainly in the field of electromagnetic compatibility, where its performance is close to that of the second-order Mur ABC.

In summary to this chapter, at the beginning of the 1990's, there existed various absorbing boundary conditions. These ABCs were highly effective as long as the waves to be absorbed were traveling waves. But they were essentially noneffective for the absorption of evanescent waves. Two novel ABCs introduced in the 1990's widely improved the simulation of free space in numerical electromagnetics. The principal reason is that these novel ABCs are able to deal with evanescent waves. These ABCs are the complementary operator method [6] and the perfectly matched layer [5].

CHAPTER 2

The Two-Dimensional Perfectly Matched Layer

In this chapter, the perfectly matched layer absorbing boundary condition is introduced in the two-dimensional (2D) case. The 2D case is simple and important because the absorption of plane waves by a PML or any other ABC is essentially a 2D problem. With a 3D computational domain, the interaction of a plane wave with a PML is a true 3D problem in the edge and corner regions, but it remains a 2D problem in the walls that form most of the outer boundary of the domain.

2.1   A MEDIUM WITHOUT REFLECTION AT NORMAL AND GRAZING INCIDENCES

Starting from medium (1.8), a new medium can be obtained by modifying the equations so as to extend the zero reflection at normal incidence to zero reflection at grazing incidence. We consider a 2D problem without field variation in z direction, in the TE case where $\boldsymbol{E}$ field is lying in the (x, y) plane and $\boldsymbol{H}$ field is parallel to z direction. In that case, Eqs. (1.8) reduce to

$$\varepsilon_0 \frac{\partial E_x}{\partial t} + \sigma E_x = \frac{\partial H_z}{\partial y} \tag{2.1a}$$

$$\varepsilon_0 \frac{\partial E_y}{\partial t} + \sigma E_y = -\frac{\partial H_z}{\partial x} \tag{2.1b}$$

$$\mu_0 \frac{\partial H_z}{\partial t} + \sigma^* H_z = \frac{\partial E_x}{\partial y} - \frac{\partial E_y}{\partial x}. \tag{2.1c}$$

Between a vacuum and this medium, the reflection coefficient is given by (1.6) with $n = 1$. The reflection is zero at normal incidence and total at grazing incidence. Let us now consider a fictitious medium obtained from (2.1) by removing conductivity σ in (2.1a) and by splitting (2.1c) into two new equations as follows:

$$\varepsilon_0 \frac{\partial E_x}{\partial t} = \frac{\partial (H_{zx} + H_{zy})}{\partial y} \qquad (2.2a)$$

$$\varepsilon_0 \frac{\partial E_y}{\partial t} + \sigma E_y = -\frac{\partial (H_{zx} + H_{zy})}{\partial x} \qquad (2.2b)$$

$$\mu_0 \frac{\partial H_{zx}}{\partial t} + \sigma^* H_{zx} = -\frac{\partial E_y}{\partial x} \qquad (2.2c)$$

$$\mu_0 \frac{\partial H_{zy}}{\partial t} = \frac{\partial E_x}{\partial y}. \qquad (2.2d)$$

The **H** field is split into two contributions, called subcomponents, denoted as H_{zx} and H_{zy}. One can consider that the two contributions are also present in (2.1) or in any physical media, but in the case of (2.2) one subcomponent, H_{zx}, is absorbed, while the other one, H_{zy}, is not absorbed.

The propagation of plane waves in medium (2.2) and the reflection from a vacuum–medium interface can be predicted without algebraic derivation in two special cases where the propagation is parallel to the x or y coordinates. To this end, let components E_y and H_z of a plane wave propagating in x direction be set as an initial condition in the medium, and let us denote H_z as H_{zx}. The space derivatives in y direction equal zero, so that the time evolution of the field in the medium is only governed by (2.2b) and (2.2c). These equations are like (2.1b) and (2.1c), so that the wave will propagate in medium (2.2) like in medium (2.1), with the same attenuation, and no E_x nor H_{zy} components will be generated. If we now consider as an initial condition a plane wave (E_x, H_{zy}) propagating in y direction, the derivatives in x direction equal zero so that the propagation is only governed by (2.2a) and (2.2d). These equations are identical to their counterparts in a vacuum. From this, the initial wave will propagate as if the medium were a vacuum, without attenuation.

Let us now consider the reflection coefficient from a vacuum–medium interface. Assume that a plane wave propagates in the vacuum in x direction and strikes the interface, that is wave A in Fig. 2.1. Since a wave without y dependence is governed in (2.2) by the same equations as in (2.1), the incident wave will penetrate into the medium without reflection. At normal incidence the reflection coefficient from medium (2.2) equals zero. Consider now an initial field equal to the field of a plane wave propagating in y direction in a vacuum. Assume this field is set on both sides of the interface, with H_z denoted as H_{zy} in the right-hand side medium, that is wave B in Fig. 2.1. In medium (2.2) this field is governed by (2.2a) and (2.2d) that are identical

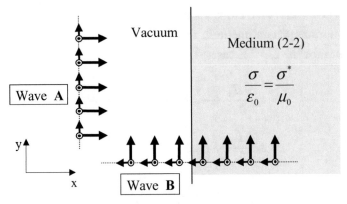

FIGURE 2.1: Plane waves at normal and grazing incidences with respect to the interface between a vacuum and medium (2.2)

to the equations in a vacuum. In consequence, wave B is governed by the same equations on the two sides of the interface. It will remain a plane wave, as if medium (2.2) were not present. No additional field will be generated on the left-hand side of the interface. Thus, the reflection coefficient from medium (2.2) equals zero at grazing incidence.

In conclusion, it can be shown without mathematics that the reflection coefficient from an interface normal to x direction, between a vacuum and medium (2.2), equals zero at two incidence angles, 0 and 90°. This is summarized in Fig. 2.2. The derivations in the next two paragraphs will show that the reflection equals zero at any incidence in the range 0–90°.

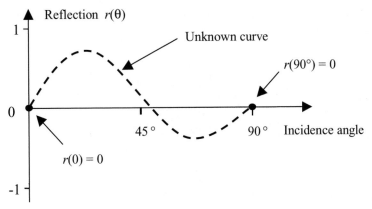

FIGURE 2.2: A simple inspection of Eqs. (2.2) shows that the reflection factor from a vacuum–medium (2.2) interface is zero at normal and grazing incidences

2.2   THE PML MEDIUM IN THE 2D TE CASE

We now consider a more general medium that holds as a special case the medium (2.2). This is the PML medium for the 2D TE case:

$$\varepsilon_0 \frac{\partial E_x}{\partial t} + \sigma_y E_x = \frac{\partial(H_{zx} + H_{zy})}{\partial y} \qquad (2.3a)$$

$$\varepsilon_0 \frac{\partial E_y}{\partial t} + \sigma_x E_y = -\frac{\partial(H_{zx} + H_{zy})}{\partial x} \qquad (2.3b)$$

$$\mu_0 \frac{\partial H_{zx}}{\partial t} + \sigma_x^* H_{zx} = -\frac{\partial E_y}{\partial x} \qquad (2.3c)$$

$$\mu_0 \frac{\partial H_{zy}}{\partial t} + \sigma_y^* H_{zy} = \frac{\partial E_x}{\partial y} \qquad (2.3d)$$

where parameters σ_x, σ_x^*, σ_y, σ_y^* are homogeneous to electric and magnetic conductivities. Medium (2.3) reduces to (2.2) in the case where $\sigma_y = \sigma_y^* = 0$. Let us now search for plane wave solutions of (2.3) of the following form:

$$E_x = E_{0x}\, e^{j\omega t - jk_x x - jk_y y} \qquad (2.4a)$$

$$E_y = E_{0y}\, e^{j\omega t - jk_x x - jk_y y} \qquad (2.4b)$$

$$H_{zx} = H_{0zx}\, e^{j\omega t - jk_x x - jk_y y} \qquad (2.4c)$$

$$H_{zy} = H_{0zy}\, e^{j\omega t - jk_x x - jk_y y} \qquad (2.4d)$$

where k_x and k_y are the components of the wave vector, and ω is the angular frequency. Inserting (2.4) into system (2.3) yields

$$\omega\varepsilon_0 \left(1 + \frac{\sigma_y}{j\omega\varepsilon_0}\right) E_{0x} = -k_y(H_{0zx} + H_{0zy}) \qquad (2.5a)$$

$$\omega\varepsilon_0 \left(1 + \frac{\sigma_x}{j\omega\varepsilon_0}\right) E_{0y} = k_x(H_{0zx} + H_{0zy}) \qquad (2.5b)$$

$$\omega\mu_0 \left(1 + \frac{\sigma_x^*}{j\omega\mu_0}\right) H_{0zx} = k_x E_{0y} \qquad (2.5c)$$

$$\omega\mu_0 \left(1 + \frac{\sigma_y^*}{j\omega\mu_0}\right) H_{0zy} = -k_y E_{0x}. \qquad (2.5d)$$

Let us now define

$$s_x = 1 + \frac{\sigma_x}{j\varepsilon_0\omega} \qquad s_x^* = 1 + \frac{\sigma_x^*}{j\mu_0\omega} \tag{2.6a}$$

$$s_y = 1 + \frac{\sigma_y}{j\varepsilon_0\omega} \qquad s_y^* = 1 + \frac{\sigma_y^*}{j\mu_0\omega} \tag{2.6b}$$

that will be called the stretching coefficients, for reasons explained later. By denoting as $H_{0z} = H_{0zx} + H_{0zy}$ the magnitude of the sum $H_z = H_{zx} + H_{zy}$, Eqs. (2.5c) and (2.5d) can merge and (2.5) can be rewritten as

$$\omega\varepsilon_0 \, E_{0x} = -\frac{k_y}{s_y} H_{0z} \tag{2.7a}$$

$$\omega\varepsilon_0 \, E_{0y} = \frac{k_x}{s_x} H_{0z} \tag{2.7b}$$

$$\omega\mu_0 \, H_{0z} = \frac{k_x}{s_x^*} E_{0y} - \frac{k_y}{s_y^*} E_{0x}. \tag{2.7c}$$

Then, by inserting E_{0x} and E_{0y} from (2.7a) and (2.7b) into (2.7c), we obtain

$$\omega^2 \varepsilon_0 \mu_0 = \frac{k_x^2}{s_x s_x^*} + \frac{k_y^2}{s_y s_y^*}. \tag{2.8}$$

Solutions of the form (2.4) can exist in the medium (2.3) provided that condition (2.8) holds. This is the equation of dispersion that connects the angular frequency ω to the possible $\mathbf{k}$ vectors in the 2D PML medium. Notice that (2.8) is like its counterpart in a vacuum, with only k_x replaced with $k_x/\sqrt{s_x s_x^*}$ and k_y replaced with $k_y/\sqrt{s_y s_y^*}$. The following wave numbers satisfy (2.8):

$$k_x = \frac{\omega}{c}\sqrt{s_x s_x^*}\cos\theta \tag{2.9a}$$

$$k_y = \frac{\omega}{c}\sqrt{s_y s_y^*}\sin\theta \tag{2.9b}$$

where θ is a free parameter and c is the speed of light. These wave numbers allow the field in the PML medium to be expressed explicitly. Inserting (2.9) into (2.4), the components of the field are of the following form, where ψ is either E_x, E_y, H_{zx}, or H_{zy}:

$$\psi = \psi_0 \, e^{j\omega t} e^{-j\frac{\omega}{c}\left[\sqrt{s_x s_x^*}\cos\theta x + \sqrt{s_y s_y^*}\sin\theta y\right]}. \tag{2.10}$$

If the two couples of conductivities (σ_x, σ_x^*) and (σ_y, σ_y^*) satisfy the matching condition (1.9), then $s_x = s_x^*$ and $s_y = s_y^*$, so that (2.10) becomes

$$\psi = \psi_0 \, e^{j\omega t} \, e^{-j\frac{\omega}{c}(x\cos\theta + y\sin\theta)} \, e^{-\frac{\sigma_x}{\varepsilon_0 c}\cos\theta x} \, e^{-\frac{\sigma_y}{\varepsilon_0 c}\sin\theta y}. \tag{2.11}$$

The first two exponentials are identical to the waveform in a vacuum. The phase of the wave propagates in direction θ with celerity c, as in a vacuum. Two absorbing terms are present. The magnitude of the wave decreases in x and y directions, according to conductivities σ_x and σ_y, respectively. If one conductivity equals zero the wave magnitude is constant in the corresponding direction. Especially, if $\sigma_y = 0$ the magnitude of the field components does not depend on the location upon a line perpendicular to x. In consequence, the variations of the phase and magnitude upon such a line in the PML are like the variations of the phase and magnitude in a vacuum. This is a necessary condition in view of removing the reflection from a vacuum–PML interface normal to x. It never holds with physical lossy media.

The components of the field can be expressed explicitly. Using (2.7) and (2.9), and with $E_0^2 = E_{0x}^2 + E_{0y}^2$, where E_0 is the E field modulus, we obtain:

$$E_{0x} = -\frac{1}{w}\sqrt{\frac{s_y^*}{s_y}}\sin\theta \, E_0 \tag{2.12a}$$

$$E_{0y} = \frac{1}{w}\sqrt{\frac{s_x^*}{s_x}}\cos\theta \, E_0 \tag{2.12b}$$

$$H_{0z} = \frac{1}{w}\sqrt{\frac{\varepsilon_0}{\mu_0}} \, E_0 \tag{2.12c}$$

$$w = \sqrt{\frac{s_x^*}{s_x}\cos^2\theta + \frac{s_y^*}{s_y}\sin^2\theta}, \tag{2.12d}$$

and in addition, using (2.5c) and (2.5d):

$$H_{0zx} = H_{0z}\cos^2\theta \tag{2.13a}$$

$$H_{0zy} = H_{0z}\sin^2\theta. \tag{2.13b}$$

Denoting as H_0 the modulus of the H field, that is $H_0 = H_{0z}$, in the case where the matching condition (1.9) holds for both (σ_x, σ_x^*) and (σ_y, σ_y^*), we have $w = 1$ and (2.12c) yields:

$$\frac{E_0}{H_0} = \sqrt{\frac{\mu_0}{\varepsilon_0}}, \tag{2.14}$$

so that the impedance of the wave is like in a vacuum, and like in the matched medium (1.8). Moreover, from (2.11), (2.12a), (2.12b), the E field is perpendicular to the direction of propagation of the phase θ, as in a vacuum. From this, at a vacuum–PML interface normal to x, if $\sigma_y = \sigma_y^* = 0$ the phase and magnitude of E and H fields in the PML can equal the phase and magnitude of E and H fields in the vacuum, provided that the angles θ in the two media are equal. Thus, in the interface a transmitted wave can match in a perfect manner any incident wave. No reflected wave is needed to ensure continuity of the tangential components of the field in the interface; in other words the reflection coefficient is zero. This is confirmed by algebraic derivations in the next paragraph.

2.3   REFLECTION OF WAVES FROM A VACUUM–PML INTERFACE AND FROM A PML–PML INTERFACE

In this paragraph the reflection coefficient is derived for an interface between two PML media. This includes the case of a vacuum–PML interface since a vacuum is nothing but a special PML medium where $\sigma_x = \sigma_x^* = \sigma_y = \sigma_y^* = 0$. An absorbing boundary condition surrounding a computational domain will be composed of vacuum–PML interfaces on the walls of the domain, and of more general PML–PML interfaces in the corners of the domain. PML–PML interfaces also will be present in actual use of the PML ABC in numerical methods where the conductivity will grow from one cell or element to the next, so that there will be inner PML–PML interfaces within the PML ABC.

Let us consider an interface between two PML media (Fig. 2.3), denoted as PML (σ_{x1}, σ_{x1}^*, σ_{y1}, σ_{y1}^*) and PML (σ_{x2}, σ_{x2}^*, σ_{y2}, σ_{y2}^*). As at an interface between physical media,

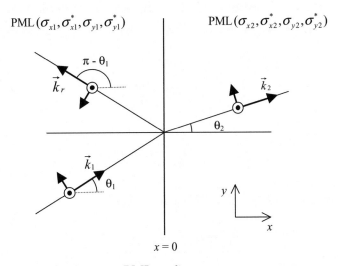

FIGURE 2.3: An interface between two PML media

the ratio of the reflected wave to the incident wave, and the ratio of the transmitted wave to the incident wave, are invariant in the interface. Using (2.4) this results in the equality of the components of the wave vectors lying in the interface, that is:

$$k_{y1} = k_{yr} = k_{y2} \qquad (2.15)$$

where k_{y1}, k_{yr}, k_{y2} are the y components of the wave vectors of the incident, reflected, and transmitted waves, respectively. For the reflected wave, using (2.9b) this yields:

$$\theta_r = \pi - \theta_1 \qquad (2.16)$$

as at any physical interface. From (2.9a) we have then $k_{xr} = -k_{x1}$. For the transmitted wave, (2.9b) and (2.15) yield

$$\sqrt{s_{y1}s_{y1}^*}\,\sin\theta_1 = \sqrt{s_{y2}s_{y2}^*}\,\sin\theta_2 \qquad (2.17)$$

Consider now the reflection and transmission between two PML media (2.3). As with physical media, components E_y and H_z lying in the interface are continuous because their space derivatives in direction x perpendicular to the interface are used in the governing equations (2.3). From this, using (2.4) and setting $x = 0$ in the interface, the following relationships hold with the magnitudes of the incident, reflected, and transmitted waves:

$$E_{0y1} + E_{0yr} = E_{0y2} \qquad (2.18a)$$

$$H_{0z1} + H_{0zr} = H_{0z2}. \qquad (2.18b)$$

Using (2.7b) for replacing H_{0z} with E_{0y} in (2.18b), defining the reflection and transmission coefficients as $r = E_{0yr}/E_{0y1}$ and $t = E_{0y2}/E_{0y1}$, and with $k_{xr} = -k_{x1}$, the set (2.18) becomes

$$1 + r = t \qquad (2.19a)$$

$$\frac{s_{x1}}{k_{x1}}(1 - r) = \frac{s_{x2}}{k_{x2}}t \qquad (2.19b)$$

from which

$$r = \frac{s_{x1}k_{x2} - s_{x2}k_{x1}}{s_{x1}k_{x2} + s_{x2}k_{x1}}. \qquad (2.20)$$

After replacement of k_{x1} and k_{x2} with (2.9), we have then

$$r = \frac{\sqrt{\frac{s_{x2}^*}{s_{x2}}}\cos\theta_2 - \sqrt{\frac{s_{x1}^*}{s_{x1}}}\cos\theta_1}{\sqrt{\frac{s_{x2}^*}{s_{x2}}}\cos\theta_2 + \sqrt{\frac{s_{x1}^*}{s_{x1}}}\cos\theta_1}. \qquad (2.21)$$

Consider the special case where the transverse conductivities σ_y and σ_y^* are equal in the two PML media, that is $s_{y1} = s_{y2}$ and $s_{y1}^* = s_{y2}^*$. Then, from (2.17), we have

$$\theta_1 = \theta_2. \tag{2.22}$$

This is the Snell–Descartes law at an interface between two PML media whose transverse conductivities are equal. If in addition we assume that the matching condition (1.9) holds in both PMLs, that is $s_x = s_x^*$, reflection (2.21) reduces to

$$r = 0. \tag{2.23}$$

Thus, at an interface normal to x between two PML media satisfying the matching condition (1.9) and whose transverse conductivities (σ_y, σ_y^*) are equal, the reflection coefficient is zero at any incidence angle and any frequency. Obviously, that is true at an interface between a vacuum and a PML $(\sigma_x, \sigma_x^*, 0, 0)$. Especially, the medium (2.2) designed so as to be reflectionless at normal and grazing incidences, as drawn in Fig. 2.2, is reflectionless at all the incidence angles as well.

In the case where the transverse conductivities of the two PML media are equal, so that (2.22) holds, if the longitudinal conductivities do not satisfy (1.9) r becomes

$$r = \frac{\sqrt{\frac{s_{x1}}{s_{x1}^*}} - \sqrt{\frac{s_{x2}}{s_{x2}^*}}}{\sqrt{\frac{s_{x1}}{s_{x1}^*}} + \sqrt{\frac{s_{x2}}{s_{x2}^*}}}. \tag{2.24}$$

The reflection does not equal zero and depends on frequency, but it does not depend on the incidence angle. This noteworthy feature has been verified by numerical experiments in [5].

In conclusion to this paragraph, there is no reflection from an interface between two matched PML media whose transverse conductivities are equal. This also holds in the important special case of an interface between a vacuum and a PML whose transverse conductivities equal zero. In the above, only the 2D TE case has been considered. In the 2D TM case, with E_z, H_x, H_y components in place of E_x, E_y, H_z, a PML can be found in the same way, by splitting E_z in place of H_z [5]. This PML also produces no reflection from vacuum–PML of PML–PML interfaces.

2.4   THE PERFECTLY MATCHED LAYER ABSORBING BOUNDARY CONDITION

The PML medium permits an absorbing boundary condition to be realized to absorb plane waves on a plane boundary. To this end, let a layer of PML medium be placed between a vacuum and a perfect electric condition (PEC), as shown in Fig. 2.4. The transverse conductivity is zero so as to cancel the reflection from the interface. An incident plane wave penetrates into the

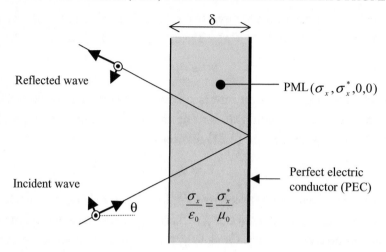

FIGURE 2.4: The PML ABC on a plane boundary

PML where its direction of propagation θ is left unchanged from (2.22). The wave is absorbed according to the real exponential depending on conductivity σ_x in (2.11). The wave is then reflected back to the vacuum from the PEC condition. Finally, the apparent reflection in the vacuum is given by the absorption corresponding to path 2δ in the PML, where δ is the PML thickness:

$$R(\theta) = e^{-2\frac{\sigma_x}{\varepsilon_0 c}\cos\theta\delta}. \tag{2.25}$$

The set (PML, PEC) in Fig. 2.4 is the PML ABC. Its reflection coefficient (2.25) does not equal zero despite the zero reflection from the vacuum–PML interface. Several remarks can be done about this reflection coefficient:

- As with previous ABCs (1.6), $R(\theta)$ tends to unity as the incidence tends to the grazing incidence. This is not a serious drawback in most problems solved with numerical methods.

- The reflection $R(\theta)$ can be lowered at will. In some way this could be thought of as being equivalent to increasing the order of analytical ABCs whose reflection is given by (1.6).

- Lowering $R(\theta)$ can be achieved either by increasing the thickness of the layer δ or by increasing the conductivity σ_x. In theory the two methods are equivalent. As will be shown and discussed later, that is not true in actual numerical methods, because sharp variations of the conductivity in a discrete space result in spurious numerical reflections. Choosing the conductivity and the thickness of the PML is a major question in applications, because contradictory requirements hold. First the thickness of the PML must be as thin as possible so as to reduce the computational cost, second the variations of the conductivity must be small enough to reduce the spurious reflection, and thirdly the theoretical reflection (2.25) must be as small as possible. In actual implementations of the PML ABC, the conductivity varies in the PML from a small value in the interface to a larger value on the outer side. Denoting as ρ the coordinate in the direction normal to the interface, and σ_ρ the conductivity in the PML, the reflection coefficient is then:

$$R(\theta) = [R(0)]^{\cos\theta} \tag{2.26a}$$

where $R(0)$ is the reflection a normal incidence:

$$R(0) = e^{-2\frac{1}{\varepsilon_0 c}\int_0^\delta \sigma_\rho(\rho)d\rho}. \tag{2.26b}$$

The choice of the profile of conductivity $\sigma(\rho)$ is discussed in detail in Chapter 6 for the application of the PML ABC to the FDTD method.

In actual problems solved by numerical methods, the boundary of the domain is not a plane, as in Fig. 2.4, but a concave surface enclosing a computational domain. The PML media allow such concave ABCs to be realized. This is depicted in Fig. 2.5. The ABC is composed of various PML media in such a way that the reflection is zero from all the inner interfaces in the domain. This is the case at the vacuum–PML interfaces, as in Fig. 2.4, where transverse conductivities equal zero on both sides of the interfaces. This is also the case at the PML–PML interfaces in the corner regions, where the transverse conductivities of the two PMLs are equal.

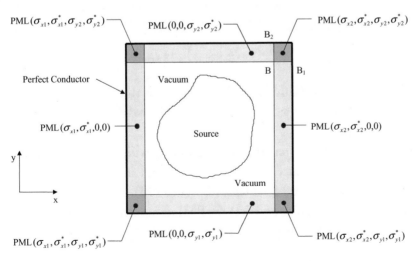

FIGURE 2.5: The PML ABC on the outer boundary of a concave volume

For instance, the transverse conductivities σ_x and σ_x^* equal σ_{x2} and σ_{x2}^*, respectively, on both sides of interface BB_1.

Since no reflection is produced from the interfaces, the reflection from the walls of the domain is given by (2.26). In the corner regions the absorption is larger because the two exponentials in (2.11) are present. In numerical methods, the PML ABC is designed by considering the absorption in the walls (2.26).

2.5   EVANESCENT WAVES IN PML MEDIA

Solution (2.9) of the wave equation (2.8) is like the solution usually considered in physical media. In fact, the more general solutions of (2.8) are of the form

$$k_x = \frac{\omega}{c}\sqrt{s_x s_x^*}\, C(\chi, \theta) \tag{2.27a}$$

$$k_y = \frac{\omega}{c}\sqrt{s_y s_y^*}\, S(\chi, \theta) \tag{2.27b}$$

where $S(\chi, \theta)$ and $C(\chi, \theta)$ are the generalized sine and cosine

$$C(\chi, \theta) = \cosh \chi \cos \theta + j \sinh \chi \sin \theta \tag{2.28a}$$

$$S(\chi, \theta) = \cosh \chi \sin \theta - j \sinh \chi \cos \theta, \tag{2.28b}$$

and χ and θ are free parameters, with $-\infty < \chi < \infty$ and $0 < \theta < 2\pi$. Notice that $C(\theta, \chi)^2 + S(\theta, \chi)^2 = 1$. Two cases are of special interest. First, if $\cosh \chi = 1$, then $C(\theta, \chi) = \cos \theta$ and $S(\theta, \chi) = \sin \theta$, so that (2.27) reduces to (2.9). This case corresponds to traveling waves propagating in direction θ with respect to the x-axis. Second, if $\cos \theta = 1$, then $C(\theta, \chi) = \cosh \chi$ and $S(\theta, \chi) = -j \sinh \chi$. By inserting k_x and k_y into (2.4), and by assuming that the matching condition (1.9) holds in the PML, the field components are of the form

$$\psi = \psi_0\, e^{j\omega\left[t - \frac{\cosh\chi}{c} x + \frac{\sigma_y}{\varepsilon_0 c\omega}\sinh\chi\, y\right]}\, e^{-\frac{\omega}{c}\sinh\chi\, y}\, e^{-\frac{\sigma_x}{\varepsilon_0 c}\cosh\chi\, x}. \tag{2.29}$$

In the special case $\sigma_x = \sigma_y = 0$, this waveform reduces to the well-known wave propagating in x direction and evanescent in y direction. In the PML, additional terms are present in the phase and in the absorption.

Let us now consider the general solution (2.27), with any χ and θ. This will yield nonuniform waves having any direction of propagation and any direction of evanescence with respect to the coordinate axes, and then with respect to PML media perpendicular either to x or y. Let us assume that (1.9) holds and that the PML is perpendicular to x with $\sigma_y = \sigma_y^* = 0$, i.e., $s_x = s_x^*$ and $s_y = s_y^* = 1$. From (2.27) the components of the wave are of the form

$$\psi = \psi_0\, e^{j\omega\left[t - \frac{\cosh\chi}{c}(x\cos\theta + y\sin\theta) - \frac{\sigma_x}{\varepsilon_0 c\omega}\sinh\chi\sin\theta\, x\right]}\, e^{-\frac{\omega}{c}\sinh\chi(y\cos\theta - x\sin\theta)}\, e^{-\frac{\sigma_x}{\varepsilon_0 c}\cosh\chi\cos\theta\, x}. \tag{2.30}$$

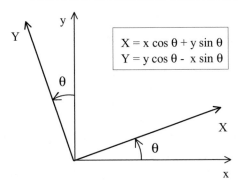

FIGURE 2.6: Direction of phase propagation X and direction of evanescence Y

Using the system of coordinates (X, Y) forming an angle θ with respect to the (x, y) system (Fig. 2.6), Eq. (2.30) can be rewritten as

$$\psi = \psi_0 \, e^{j\omega\left[t-\frac{\cosh\chi}{c}X-\frac{\sigma_x}{\varepsilon_0 c\omega}\sinh\chi\sin\theta x\right]} e^{-\frac{\omega}{c}\sinh\chi Y} e^{-\frac{\sigma_x}{\varepsilon_0 c}\cosh\chi\cos\theta x}. \qquad (2.31)$$

In (2.31) the exponential terms depending on X and Y are exactly the waveform of a wave propagating in X direction and evanescent in Y direction in a vacuum. Two additional terms depending on x and σ_x are present: first a phase term, second an absorbing term. It is worth noticing from (2.31) that:

- The evanescent waves are absorbed in the PML medium, due to the last exponential in (2.31). The absorption is larger than that of traveling waves, because $\cosh\chi > 1$ for evanescent waves instead of $\cosh\chi = 1$ for traveling waves. As will be discussed later, this may result in a spurious reflection from PMLs in numerical methods, because for strongly evanescent waves ($\cosh\chi \gg 1$) the absorption is enormous.

- The absorption of evanescent waves whose direction of propagation is parallel to the interface and direction of evanescence is perpendicular to the interface, that is $\theta = \pm \pi/2$ in (2.31), are not absorbed. This is a drawback in some applications, as will be illustrated in Chapter 6 with waveguide problems.

- The phase of evanescent waves depends on frequency, due to the term depending on σ_x in the first exponential of (2.31). For strongly evanescent waves ($\sinh\chi \gg 1$) the phase may rapidly vary with distance at low frequency. Again, this may result in spurious reflections in numerical methods.

Let us now consider the components of the field in the PML medium. They are identical to (2.12) and (2.13), with only the replacement of $\cos\theta$ and $\sin\theta$ with $C(\chi, \theta)$ and $S(\chi, \theta)$. If the matching condition (1.9) holds, in the (X, Y) coordinates defined in Fig. 2.6 the H field (2.12c) is left unchanged while the E components become:

$$E_{0X} = j \sinh \chi \, E_0 \qquad (2.32a)$$

$$E_{0Y} = \cosh \chi \, E_0, \qquad (2.32b)$$

that are nothing but the components of a wave propagating in X direction and evanescent in Y direction in a vacuum. Thus, in a PML perpendicular to x, i.e., if $\sigma_y = 0$, evanescent waves are like in a vacuum, with only the addition of phase and absorbing terms depending on x.

Consider now an interface between two PML media, with a nonuniform wave propagating from medium 1 toward medium 2. We assume in the following that the matching condition (1.9) holds in both two media. The components of the wave vectors of the incident, reflected, and transmitted waves are equal in the interface, i.e., (2.15) holds. Using (2.27b) this yields $S(\chi_r, \theta_r) = S(\chi_1, \theta_1)$, and if in addition the transverse conductivities σ_y are equal, $S(\chi_2, \theta_2) = S(\chi_1, \theta_1)$, where $\chi_r, \theta_r, \chi_2, \theta_2$ are the unknown χ and θ of the reflected and transmitted waves. By solving for the unknowns χ and θ the equation $S(\chi, \theta) = S(\chi_1, \theta_1)$, which is equivalent to two real equations (real and imaginary parts), the following two solutions corresponding to the reflected and transmitted waves are obtained:

$$\chi_r = -\chi_1 \quad \theta_r = \pi - \theta_1 \qquad (2.33a)$$

$$\chi_2 = \chi_1 \quad \theta_2 = \theta_1. \qquad (2.33b)$$

These equations are the generalization of (2.16) and (2.22) to nonhomogeneous waves. From (2.27a) and (2.33a) we have $k_{xr} = -k_{x1}$, as with traveling waves. From (2.33b) the evanescence coefficient $\cosh\chi$ and the direction of propagation θ are left unchanged through PML–PML or vacuum–PML interfaces.

The reflection coefficient can be found by enforcing the continuity of components E_y and H_z lying in the interface. This yields (2.18). Using (2.7b), Eqs. (2.19) and (2.20) are obtained. Then, with (2.27a) reflection (2.21) is obtained with $C(\chi_1, \theta_1)$ and $C(\chi_2, \theta_2)$ in place of $\cos\theta_1$ and $\cos\theta_2$. Finally, provided that (1.9) holds and the transverse conductivities are equal, so that (2.33b) holds, we have

$$r = 0. \qquad (2.34)$$

In summary to this paragraph, evanescent waves are not reflected from the interface between a vacuum and a PML medium, or more generally from the interface between two PML media whose transverse conductivities are equal. In the PML such waves are absorbed according to the coefficient in (2.30) and (2.31). This PML attenuation is added to the natural decrease of evanescent waves. Finally, in a PML we can write:

$$|\psi_{\text{PML}}| = |\psi_{\text{vacuum}}| \, e^{-\frac{\sigma_x}{\varepsilon_0 c} \cosh \chi \cos \theta x}, \qquad (2.35)$$

where ψ_{vacuum} is the waveform in a vacuum. The absorption is larger than the absorption of purely traveling waves that correspond to the special case $\cosh \chi = 1$ in (2.35). The coefficient in (2.35) is of primary importance for the interpretation of the spurious reflection from PML ABCs in numerical methods [15].

CHAPTER 3

Generalizations and Interpretations of the Perfectly Matched Layer

In this chapter, the perfectly matched layer absorbing boundary condition is generalized to three dimensions and to more general physical media. The first part is devoted to the 3D PML in the case where the inner medium is filled with a vacuum. The second part presents two interpretations of the PML medium, in terms of stretched coordinates and in terms of dependent currents. In the third part the PML is generalized to other physical media, especially to anisotropic media. The fourth part addresses the case where the inner medium is nonhomogeneous. In the fifth part the absorbing medium called uniaxial PML is presented. Finally, the sixth part introduces the CFS-PML medium obtained from the normal PML medium by means of a modification of the stretching coefficient.

3.1   THE THREE-DIMENSIONAL PML MATCHED TO A VACUUM

As discussed in the previous chapter about the 2D case, the absence of reflection from a vacuum–PML interface is closely related to the fact that plane waves in the PML can propagate without attenuation in the direction parallel to the interface. In 3D, a similar property can be obtained by a straightforward generalization of Eqs. (2.3). The six components of the field are split into 12 subcomponents and the six Cartesian equations are split into 12 subequations, with two independent conductivities assigned to the two subequations resulting from each equation:

$$\varepsilon_0 \frac{\partial E_{xy}}{\partial t} + \sigma_y E_{xy} = \frac{\partial (H_{zx} + H_{zy})}{\partial y} \tag{3.1a}$$

$$\varepsilon_0 \frac{\partial E_{xz}}{\partial t} + \sigma_z E_{xz} = -\frac{\partial (H_{yz} + H_{yx})}{\partial z} \tag{3.1b}$$

$$\varepsilon_0 \frac{\partial E_{yz}}{\partial t} + \sigma_z E_{yz} = \frac{\partial (H_{xy} + H_{xz})}{\partial z} \tag{3.1c}$$

$$\varepsilon_0 \frac{\partial E_{yx}}{\partial t} + \sigma_x E_{yx} = -\frac{\partial (H_{zx} + H_{zy})}{\partial x} \tag{3.1d}$$

$$\varepsilon_0 \frac{\partial E_{zx}}{\partial t} + \sigma_x E_{zx} = \frac{\partial (H_{yz} + H_{yx})}{\partial x} \tag{3.1e}$$

$$\varepsilon_0 \frac{\partial E_{zy}}{\partial t} + \sigma_y E_{zy} = -\frac{\partial (H_{xy} + H_{xz})}{\partial y} \tag{3.1f}$$

$$\mu_0 \frac{\partial H_{xy}}{\partial t} + \sigma_y^* H_{xy} = -\frac{\partial (E_{zx} + E_{zy})}{\partial y} \tag{3.1g}$$

$$\mu_0 \frac{\partial H_{xz}}{\partial t} + \sigma_z^* H_{xz} = \frac{\partial (E_{yz} + E_{yx})}{\partial z} \tag{3.1h}$$

$$\mu_0 \frac{\partial H_{yz}}{\partial t} + \sigma_z^* H_{yz} = -\frac{\partial (E_{xy} + E_{xz})}{\partial z} \tag{3.1i}$$

$$\mu_0 \frac{\partial H_{yx}}{\partial t} + \sigma_x^* H_{yx} = \frac{\partial (E_{zx} + E_{zy})}{\partial x} \tag{3.1j}$$

$$\mu_0 \frac{\partial H_{zx}}{\partial t} + \sigma_x^* H_{zx} = -\frac{\partial (E_{yx} + E_{yz})}{\partial x} \tag{3.1k}$$

$$\mu_0 \frac{\partial H_{zy}}{\partial t} + \sigma_y^* H_{zy} = \frac{\partial (E_{xy} + E_{xz})}{\partial y}. \tag{3.1l}$$

By inserting a plane wave solution in (3.1), with subcomponents of the form

$$\psi = \psi_0 \, e^{j\omega t} e^{-j k_x x - j k_y y - j k_z z}, \tag{3.2}$$

and by defining s_u and s_u^* as

$$s_u = 1 + \frac{\sigma_u}{j\omega\varepsilon_0}; \quad s_u^* = 1 + \frac{\sigma_u^*}{j\omega\mu_0} \quad (u = x, y, z), \tag{3.3}$$

we obtain 12 equations connecting the angular frequency ω, the wave numbers k_x, k_y, k_z, and the magnitudes of the subcomponents E_{0xy}, E_{0xz}, ..., H_{0zx}, H_{0zy}. For example, the first two Eqs. (3.1a) and (3.1b) yield

$$\varepsilon_0 \, s_y \, E_{0xy} = -jk_y(H_{0zx} + H_{0zy}) \tag{3.4a}$$

$$\varepsilon_0 \, s_z \, E_{0xz} = jk_z(H_{0yz} + H_{0yx}). \tag{3.4b}$$

Dividing (3.4a) and (3.4b) with s_y and s_z, respectively, and with $E_{0x} = E_{0xy} + E_{0xz}$, $H_{0z} = H_{0zx} + H_{0zy}$, and $H_{0y} = H_{0yz} + H_{0yx}$, the two Eqs. (3.4) can be merged into a single equation. By proceeding similarly with the other five couples of equations derived from (3.1c)–(3.1l), the following set is obtained:

$$\omega \varepsilon_0 \, E_{0x} = -\frac{k_y}{s_y} H_{0z} + \frac{k_z}{s_z} H_{0y} \tag{3.5a}$$

$$\omega \varepsilon_0 \, E_{0y} = -\frac{k_z}{s_z} H_{0x} + \frac{k_x}{s_x} H_{0z} \tag{3.5b}$$

$$\omega \varepsilon_0 \, E_{0z} = -\frac{k_x}{s_x} H_{0y} + \frac{k_y}{s_y} H_{0x} \tag{3.5c}$$

$$\omega \mu_0 \, H_{0x} = \frac{k_y}{s_y^*} E_{0z} - \frac{k_z}{s_z^*} E_{0y} \tag{3.5d}$$

$$\omega \mu_0 \, H_{0y} = \frac{k_z}{s_z^*} E_{0x} - \frac{k_x}{s_x^*} E_{0z} \tag{3.5e}$$

$$\omega \mu_0 \, H_{0z} = \frac{k_x}{s_x^*} E_{0y} - \frac{k_y}{s_y^*} E_{0x}. \tag{3.5f}$$

Let us now define two vectors k_s and k_s^* as follows:

$$\overrightarrow{k_s} = (k_x/s_x, k_y/s_y, k_z/s_z)^T \tag{3.6a}$$

$$\overrightarrow{k_s^*} = (k_x/s_x^*, k_y/s_y^*, k_z/s_z^*)^T. \tag{3.6b}$$

Then, (3.5) can be rewritten as

$$\varepsilon_0 \omega \overrightarrow{E_0} = -\overrightarrow{k_s} \times \overrightarrow{H_0} \tag{3.7a}$$

$$\mu_0 \omega \overrightarrow{H_0} = \overrightarrow{k_s^*} \times \overrightarrow{E_0} \tag{3.7b}$$

where E_0 and H_0 are vectors of components E_{0x}, E_{0y}, E_{0z}, and H_{0x}, H_{0y}, H_{0z}, respectively. System (3.7) is like its counterpart in a vacuum, with vectors (3.6) in place of the wave vector k. Notice that E and H fields are perpendicular from (3.7), as in a vacuum. Using H_0 from (3.7b) into (3.7a) yields

$$\varepsilon_0 \mu_0 \omega^2 \overrightarrow{E_0} = -\overrightarrow{k_s} \times (\overrightarrow{k_s^*} \times \overrightarrow{E_0}) \tag{3.8}$$

that can be rewritten as

$$\varepsilon_0 \mu_0 \omega^2 \overrightarrow{E_0} = -\left(\overrightarrow{k_s} \cdot \overrightarrow{E_0}\right) \cdot \overrightarrow{k_s^*} + \left(\overrightarrow{k_s} \cdot \overrightarrow{k_s^*}\right) \cdot \overrightarrow{E_0}. \tag{3.9}$$

From (3.7a) k_s is perpendicular to E field, so that

$$\left(\varepsilon_0 \mu_0 \omega^2 - \overrightarrow{k_s} \cdot \overrightarrow{k_s^*}\right) \overrightarrow{E_0} = 0. \tag{3.10}$$

This yields the following equation of dispersion,

$$\varepsilon_0 \mu_0 \omega^2 = \frac{k_x^2}{s_x s_x^*} + \frac{k_y^2}{s_y s_y^*} + \frac{k_z^2}{s_z s_z^*}, \tag{3.11}$$

which is the generalization of (2.8). Wave numbers satisfying (3.11) can be written in the form:

$$k_x = \frac{\omega}{c}\sqrt{s_x s_x^*} \sin\theta \cos\varphi \tag{3.12a}$$

$$k_y = \frac{\omega}{c}\sqrt{s_y s_y^*} \sin\theta \sin\varphi \tag{3.12b}$$

$$k_z = \frac{\omega}{c}\sqrt{s_z s_z^*} \cos\theta \tag{3.12c}$$

where θ and φ are free parameters.

Let us now assume that the matching condition (1.9) holds for the three couples of conductivities of the PML medium. Then, from (3.3) and (3.6) we have $\boldsymbol{k}_s = \boldsymbol{k}_s^*$, and from (3.7) $\boldsymbol{k}_s$ is perpendicular to $\boldsymbol{E}$ and $\boldsymbol{H}$. Moreover, using (3.11) and (3.7a) or (3.7b) leads to (2.14), i.e., the impedance is matched to that of a vacuum. And finally, (3.12) shows that $\boldsymbol{k}_s$ is oriented in direction (θ, φ), so that the $(\boldsymbol{E}, \boldsymbol{H})$ plane is perpendicular to direction (θ, φ).

Using now (3.2) and (3.12), any component or subcomponent is of the form

$$\psi = \psi_0\, e^{j\omega t}\, e^{-j\frac{\omega}{c}(x\sin\theta\cos\varphi + y\sin\theta\sin\varphi + z\cos\theta)}\, e^{-\frac{\sigma_x}{\varepsilon_0 c}\sin\theta\cos\varphi x}\, e^{-\frac{\sigma_y}{\varepsilon_0 c}\sin\theta\sin\varphi y}\, e^{-\frac{\sigma_z}{\varepsilon_0 c}\cos\theta z}. \tag{3.13}$$

From this, the phase propagates in direction (θ, φ) with celerity c, as in a vacuum. Denoting as η_x, η_y, η_z, the angles that direction (θ, φ) forms with the axes of coordinates, (3.13) can be rewritten in a more symmetric form with respect to the directions of space:

$$\psi = \psi_0\, e^{j\omega t}\, e^{-j\frac{\omega}{c}(x\cos\eta_x + y\sin\eta_y + z\cos\eta_z)}\, e^{-\frac{\sigma_x}{\varepsilon_0 c}\cos\eta_x x}\, e^{-\frac{\sigma_y}{\varepsilon_0 c}\cos\eta_y y}\, e^{-\frac{\sigma_z}{\varepsilon_0 c}\cos\eta_z z}. \tag{3.14}$$

Equations (3.13) or (3.14) show that the attenuation in the 3D PML can be controlled at will in the three directions. As an example, the attenuation in x and y directions can be vanished by choosing $\sigma_x = \sigma_y = 0$. In such a case the phase propagates as in a vacuum, $\boldsymbol{E}$ and $\boldsymbol{H}$ fields are perpendicular to the direction of propagation of the phase, and the wave is only attenuated in z direction. In that case, at a vacuum–PML interface perpendicular to z the transmitted wave can be perfectly matched to the incident wave, so that no reflected wave is needed to satisfy continuity of $\boldsymbol{E}$ and $\boldsymbol{H}$ fields in the interface.

To prove mathematically the absence of reflection from a PML–PML interface, let us consider an interface perpendicular to z. The equality of the wave numbers in the interface can be expressed as:

$$k_{x1} = k_{xr} = k_{x2} \tag{3.15a}$$

$$k_{y1} = k_{yr} = k_{y2} \tag{3.15b}$$

where indexes 1, r, and 2 denote the incident, reflected, and transmitted waves, respectively. From (3.12a), (3.12b), (3.15), the direction of the reflected wave ($\theta > \pi/2$) is

$$\varphi_r = \varphi_1 \quad \text{and} \quad \theta_r = \pi - \theta_1 \tag{3.16}$$

as at any physical interface. In addition, from (3.12c) we have $k_{zr} = -k_{z1}$. For the wave transmitted into the PML ($\theta < \pi/2$), (3.12a), (3.12b), (3.15) yield

$$\sqrt{s_{x1}s_{x1}^*} \sin\theta_1 \cos\varphi_1 = \sqrt{s_{x2}s_{x2}^*} \sin\theta_2 \cos\varphi_2 \tag{3.17a}$$

$$\sqrt{s_{y1}s_{y1}^*} \sin\theta_1 \sin\varphi_1 = \sqrt{s_{y2}s_{y2}^*} \sin\theta_2 \sin\varphi_2. \tag{3.17b}$$

Assume now that the incident field is polarized in such a way that the E field is perpendicular to z direction (E parallel to the interface). Only components E_x and E_y are present. Continuity of the field components lying in the interface ($x = 0$) yields

$$E_{0x1} + E_{0xr} = E_{0x2} \tag{3.18a}$$

$$E_{0y1} + E_{0yr} = E_{0y2} \tag{3.18b}$$

$$H_{0x1} + H_{0xr} = H_{0x2} \tag{3.18c}$$

$$H_{0y1} + H_{0yr} = H_{0y2}. \tag{3.18d}$$

Moreover, using (3.7b) and $E_{0z} = 0$ we have in both media:

$$H_{0x} = \frac{-k_{sz}^* E_{0y}}{\mu_0 \omega} \tag{3.19a}$$

$$H_{0y} = \frac{k_{sz}^* E_{0x}}{\mu_0 \omega}. \tag{3.19b}$$

This allows the H field to be replaced with the E field in (3.18c) and (3.18d). Using then (3.6) and $k_{zr} = -k_{z1}$, Eqs. (3.18c) and (3.18d) become

$$\frac{k_{z1}}{s_{z1}^*} E_{0y1} - \frac{k_{z1}}{s_{z1}^*} E_{0yr} = \frac{k_{z2}}{s_{z2}^*} E_{0y2} \tag{3.20a}$$

$$\frac{k_{z1}}{s_{z1}^*} E_{0x1} - \frac{k_{z1}}{s_{z1}^*} E_{0xr} = \frac{k_{z2}}{s_{z2}^*} E_{0x2}. \tag{3.20b}$$

If we define the reflection and transmission coefficients as $r_x = E_{0xr}/E_{0x1}$, $r_y = E_{0yr}/E_{0y1}$, $t_x = E_{0x2}/E_{0x1}$, $t_y = E_{0y2}/E_{0y1}$, Eqs. (3.18a), (3.18b), (3.20) can be rewritten as

$$1 + r_x = t_x \qquad (3.21a)$$

$$1 + r_y = t_y \qquad (3.21b)$$

$$k_{z1}s_{z2}^*(1 - r_x) = k_{z2}s_{z1}^* t_x \qquad (3.21c)$$

$$k_{z1}s_{z2}^*(1 - r_y) = k_{z2}s_{z1}^* t_y. \qquad (3.21d)$$

This system yields $r_x = r_y$ and $t_x = t_y$. Denoting then as r_s the common value of the reflection, i.e., $r_s = r_x = r_y$, we have

$$r_s = \frac{k_{z1}s_{z2}^* - k_{z2}s_{z1}^*}{k_{z1}s_{z2}^* + k_{z2}s_{z1}^*}, \qquad (3.22)$$

and with (3.12c):

$$r_s = \frac{\sqrt{s_{z1}/s_{z1}^*}\cos\theta_1 - \sqrt{s_{z2}/s_{z2}^*}\cos\theta_2}{\sqrt{s_{z1}/s_{z1}^*}\cos\theta_1 + \sqrt{s_{z2}/s_{z2}^*}\cos\theta_2}. \qquad (3.23)$$

Assume now that the transverse conductivities σ_x and σ_y are equal on the two sides of the interface. Then, (3.17) reduces to

$$\varphi_1 = \varphi_2 \quad \text{and} \quad \theta_1 = \theta_2. \qquad (3.24)$$

The direction of propagation of the phase is left unchanged at the interface between two PML media having the same transverse losses. If in addition the matching condition (1.9) holds for the longitudinal conductivity σ_z, then (3.23) becomes

$$r_s = 0. \qquad (3.25)$$

If we now consider the polarization with $\boldsymbol{H}$ field perpendicular to z direction ($\boldsymbol{H}$ parallel to the interface), a derivation similar to the previous one yields the following reflection coefficient in place of (3.22):

$$r_p = \frac{k_{z2}s_{z1} - k_{z1}s_{z2}}{k_{z2}s_{z1} + k_{z1}s_{z2}}. \qquad (3.26)$$

Notice that (3.26) is identical to (2.20), because the $\boldsymbol{H}$ field is also parallel to the interface in the 2D TE case. In the case of matched PML media with the same transverse conductivities, using (3.12) and (3.17) in (3.26) also leads to zero reflection from 3D interfaces, that is $r_p = 0$. Finally,

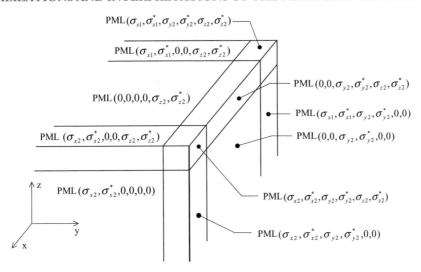

FIGURE 3.1: The three-dimensional PML absorbing boundary condition

since any polarization of the incident wave can be split into two nonparallel polarizations, $r = 0$ for any incident wave.

In summary, in the general 3D case no reflection is produced from the interface between two PML media whose transverse conductivities are equal and whose conductivities satisfy the matching condition (1.9).

3.2   THE THREE-DIMENSIONAL PML ABSORBING BOUNDARY CONDITION

The 3D PML ABC is a trivial generalization of the 2D one. In the PML matched to a vacuum, the transverses conductivities equal zero and (1.9) holds for the longitudinal conductivities. The PML is bounded with a PEC condition, as in Fig. 2.4. Denoting as ρ the direction perpendicular to the interface and σ_ρ the conductivity in this direction, from (3.14) the reflection from a PML of thickness δ with a conductivity depending on ρ is given by (2.26).

On the outer boundary of a concave computational domain, PML media with transverse conductivities equal to zero are set on the six walls, so as to cancel the reflection from the vacuum–PML interfaces. In the edge and corner regions, the conductivities are chosen as depicted in Fig. 3.1. This allows the transverse conductivities to be equal at all the PML–PML interfaces, so that no reflection is produced from these inner interfaces. For example, in the corner shown in Fig. 3.1, through the interface between the PML $(\sigma_{x2}, \sigma_{x2}^*, 0, 0, \sigma_{z2}, \sigma_{z2}^*)$ and the PML $(\sigma_{x2}, \sigma_{x2}^*, \sigma_{y2}, \sigma_{y2}^*, \sigma_{z2}, \sigma_{z2}^*)$, the transverse conductivities σ_x and σ_z are equal, ensuring then zero reflection.

3.3   INTERPRETATION OF THE PML MEDIUM IN TERMS OF STRETCHED COORDINATES

Different interpretations and formulations of PML media have been presented and discussed in the literature. The most important and useful is the interpretation in terms of stretched coordinates. This interpretation facilitates the extension of the PML concept to more general media and to other partial differential equations of physics.

Let us consider subequations (3.1a) and (3.1b). By replacing the time derivatives with $j\omega$ and by adding these two subequations, a unique equation is obtained, with components E_x, H_y, H_z in place of the subcomponents. Merging similarly the other five sets of subequations results in the following six equations that are equivalent to (3.1) for the considered angular frequency ω:

$$j\omega \, \varepsilon_0 \, E_x = \frac{1}{s_y}\frac{\partial H_z}{\partial y} - \frac{1}{s_z}\frac{\partial H_y}{\partial z} \qquad (3.27a)$$

$$j\omega \, \varepsilon_0 \, E_y = \frac{1}{s_z}\frac{\partial H_x}{\partial z} - \frac{1}{s_x}\frac{\partial H_z}{\partial x} \qquad (3.27b)$$

$$j\omega \, \varepsilon_0 \, E_z = \frac{1}{s_x}\frac{\partial H_y}{\partial x} - \frac{1}{s_y}\frac{\partial H_x}{\partial y} \qquad (3.27c)$$

$$j\omega \, \mu_0 \, H_x = -\frac{1}{s_y^*}\frac{\partial E_z}{\partial y} + \frac{1}{s_z^*}\frac{\partial E_y}{\partial z} \qquad (3.27d)$$

$$j\omega \, \mu_0 \, H_y = -\frac{1}{s_z^*}\frac{\partial E_x}{\partial z} + \frac{1}{s_x^*}\frac{\partial E_z}{\partial x} \qquad (3.27e)$$

$$j\omega \, \mu_0 \, H_z = -\frac{1}{s_x^*}\frac{\partial E_y}{\partial x} + \frac{1}{s_y^*}\frac{\partial E_x}{\partial y} \qquad (3.27f)$$

where s and s^* parameters are coefficients (3.3) and field components are the sum of the corresponding subcomponents in (3.1), for instance $E_x = E_{xy} + E_{xz}$ and $H_z = H_{zx} + H_{zy}$. Consider now coefficients s_y and s_z that may vary with y and z, and let the following change of variables be defined as

$$dy' = s_y(y)dy \quad dz' = s_z(z)dz, \qquad (3.28)$$

or equivalently:

$$y' = \int_{y_1}^{y_2} s_y(y)dy \quad z' = \int_{z_1}^{z_2} s_z(z)dz \qquad (3.29)$$

that corresponds to a stretch of coordinates, where y_1, y_2, z_1, z_2 are the limits of the stretched region. Then, Eq. (3.27a) can be rewritten as:

$$j\omega\varepsilon_0 E_x = \frac{\partial H_z}{\partial y'} - \frac{\partial H_y}{\partial z'} \qquad (3.30)$$

which is nothing but the x component of the Maxwell–Ampere equation in a vacuum, in the stretched coordinates defined with (3.28). Applying the same procedure to the other five equations of (3.27), the PML medium can be regarded as a vacuum where the coordinates are stretched with the complex factors (3.3). In the general case, the stretching factors are different for the Maxwell–Ampere and Maxwell–Faraday equations. If the PML medium is matched the stretching factors are the same for the two equations. The stretching factors of the three directions of space can be different. Especially, if two conductivities equal zero ($\sigma_x = \sigma_y = 0$ for example), only the third direction is stretched (z in this example). This is the case in the walls of a computational domain where the PML is equivalent to a stretch of the coordinate perpendicular to the interface. Notice that the stretch of distances is consistent with Eqs. (3.5). Here the wave numbers are stretched with the inverse of coefficients (3.3). This corresponds to a stretch of wavelengths with factors (3.3).

With the interpretation in terms of stretched coordinates, the required condition for the absence of reflection from a vacuum–PML can be reformulated. The coordinates must only be stretched in the direction perpendicular to the interface. More generally, at a PML–PML interface the transverse stretches must be identical on the two sides of the interface, an extra stretch can only be added in the direction normal to the interface.

The interpretation of the PML medium in terms of stretched coordinates was initially presented in [16], [17], and [18]. In [16], the PML medium is described by a formalism derived from this interpretation.

3.4   INTERPRETATION IN TERMS OF DEPENDENT CURRENTS

The time domain equations of the PML (3.1) are known as the split equations or the split PML, because these 12 equations are obtained by splitting the six components of the Maxwell equations. Other formulations without splitting have been found, such as the unsplit formulation [18, 19]. To derive this formulation, let us rewrite for example (3.27b) as

$$j\omega\varepsilon_0 s_z s_x E_y = s_x \frac{\partial H_x}{\partial z} - s_z \frac{\partial H_z}{\partial x}. \qquad (3.31)$$

In the special case where $\sigma_y = \sigma_z = 0$ (wall PML perpendicular to x) this yields in time domain:

$$\varepsilon_0 \frac{\partial E_y}{\partial t} + \sigma_x E_y = \frac{\partial H_x}{\partial z} - \frac{\partial H_z}{\partial x} + \frac{\sigma_x}{\varepsilon_0} \frac{\partial}{\partial z} \int H_x \, dt'. \qquad (3.32)$$

Similar equations can be derived from (3.27) for the time derivatives of E_z, H_y, H_z. The equations for E_x and H_x are like in a vacuum. With time domain numerical methods, these equations can be used for the advance on time of the six components of the field in the PML.

From a physical point of view, Eq. (3.32) shows that a wall PML medium—having only one nonzero conductivity—can be regarded as a physical lossy medium with dependent electric and magnetic current densities [19] proportional to the integral of the longitudinal components of the field (E_x and H_x in a PML perpendicular to x). This allows computational codes to deal with the six components of the fields, but with two additional quantities corresponding to integrals like the one in (3.32).

3.5   THE PML MATCHED TO GENERAL MEDIA

In the previous chapters, we only considered PML media matched to a vacuum. These PMLs can only be used to absorb outgoing waves on the boundary of computational domains surrounded with a vacuum. This is the case in many applications of computational electromagnetics, but in some other cases it is worth setting the PML within a medium that may be lossy or anisotropic. In the following, the PML medium is generalized to PML media matched to any physical media. This is done in frequency domain. The corresponding time domain equations are derived in the next chapter.

A trivial generalization is the PML matched to dielectric media. In that case, by replacing ε_0 with ε in the previous paragraphs, the derivations are left unchanged and the conclusions are the same. No reflection is produced from the interface between a dielectric of permeability ε and a PML governed by Eqs. (3.1) or (3.7) with ε_0 replaced with ε. This also holds if μ_0 is replaced with μ. In the case where ε_0 or μ_0 are replaced with complex numbers ε or μ in (3.7), the derivations following (3.7) remain valid. From this, the frequency domain equations of a PML matched to any isotropic medium can be obtained by stretching the coordinates, as in the case of a vacuum.

In the case of anisotropic media, either lossy or not, a frequency-domain PML can also be obtained by stretching the coordinates. This can be proved, rigorously. To this end, let us consider an electrically and magnetically anisotropic medium where permittivity ε and permeability μ are real or complex tensors. In frequency and wave vector domains, the Maxwell equations read

$$\omega \overline{\overline{\varepsilon}} \vec{E_0} = -\vec{k} \times \vec{H_0} \qquad (3.33a)$$

$$\omega \overline{\overline{\mu}} \vec{H_0} = \vec{k} \times \vec{E_0}. \qquad (3.33b)$$

Assume now that the coordinates are stretched by factors like (3.3). As in a vacuum, the stretch consists in using transformation (3.28) in the derivatives of the curls of the Maxwell equations.

We assume that the electric and magnetic stretches are equal, that is the matching condition (1.9) holds. This results in the following equations that replace (3.33), with $\boldsymbol{k}_s$ vector (3.6) in place of $\boldsymbol{k}$ vector:

$$\omega \overline{\overline{\varepsilon}} \overrightarrow{E_0} = -\overrightarrow{k_s} \times \overrightarrow{H_0} \qquad (3.34a)$$

$$\omega \overline{\overline{\mu}} \overrightarrow{H_0} = \overrightarrow{k_s} \times \overrightarrow{E_0}. \qquad (3.34b)$$

The equation of dispersion obtained by combining the two Eqs. (3.34) is identical to that obtained from Eqs. (3.33), with $\boldsymbol{k}_s$ in place of $\boldsymbol{k}$. From this, $\boldsymbol{k}_s$ in any stretched medium is identical to $\boldsymbol{k}$ in the corresponding nonstretched medium. This can be written as:

$$\overrightarrow{k_s}\Big|_{\text{PML}} \equiv \overrightarrow{k}\Big|_{\text{medium}} \qquad (3.35)$$

where $\equiv$ means that the two vectors are solutions of the same equation of dispersion. In the special case where the PML is only stretched in z direction, we have

$$k_x|_{\text{PML}} \equiv k_x|_{\text{medium}} \qquad (3.36a)$$

$$k_y|_{\text{PML}} \equiv k_y|_{\text{medium}} \qquad (3.36b)$$

$$k_z|_{\text{PML}} \equiv s_z \, k_z|_{\text{medium}}. \qquad (3.36c)$$

The phase and attenuation in x and y directions are like in the anisotropic medium. Conversely, the complex stretch s_z attenuates the wave in z direction. This attenuation is added to the physical attenuation if the anisotropic medium is lossy.

Consider now the reflection from a medium–PML interface perpendicular to z, with stretch only in z direction within the PML. The equality of transverse wave numbers (3.15) also holds, so that $k_{1x} = k_{2x} = k_{s2x}$ and $k_{1y} = k_{2y} = k_{s2y}$, where $\boldsymbol{k}_1$ is the wave vector of the incident wave in the anisotropic medium, and $\boldsymbol{k}_{s2}$ is the $\boldsymbol{k}_s$ vector of the wave transmitted into the PML. Since $\boldsymbol{k}_1$ and $\boldsymbol{k}_{s2}$ are solutions of the same equation of dispersion, as expressed by (3.35), their third components k_{1z} and k_{s2z} are equal. Then, we have

$$\overrightarrow{k_{s2}} = \overrightarrow{k_1}. \qquad (3.37)$$

The transverse fields are continuous across the interface, so that in the case where the incident $\boldsymbol{E}$ field is perpendicular to z direction, set (3.18) also holds. To derive the reflection coefficient, $\boldsymbol{H}$ field is replaced with $\boldsymbol{E}$ field in (3.18c) and (18.d) by using the Maxwell equations in the two media. Consider for instance (3.18c). Using (3.33b) and (3.34b), and replacing $\boldsymbol{k}_{s2}$ with $\boldsymbol{k}_1$ from (3.37) on the right-hand side, we obtain:

$$\left[\overline{\overline{\mu}}^{-1} \overrightarrow{k_1} \times \overrightarrow{E_{01}} \right]_x + \left[\overline{\overline{\mu}}^{-1} \overrightarrow{k_r} \times \overrightarrow{E_{0r}} \right]_x = \left[\overline{\overline{\mu}}^{-1} \overrightarrow{k_1} \times \overrightarrow{E_{02}} \right]_x \qquad (3.38)$$

where $\boldsymbol{k}_r$ is the wave vector of the reflected wave. This equation is nothing but the equation that would be obtained if the media on the two sides of the interface were identical. This also holds with the equation derived from (3.18d). Finally, the system (3.18) that gives the reflected field $\boldsymbol{E}_{0r}$ is like at an interface between two identical media. From this, for the considered polarization, there is no reflection between any anisotropic medium and its PML counterpart obtained by stretching the coordinate perpendicular to the interface. The same conclusion also holds with the other polarization, where the $\boldsymbol{H}$ field is perpendicular to z.

Obviously, the absence of reflection also holds in the more general case of an interface between two PML media obtained by stretching the same anisotropic medium, provided that the transverse stretches of coordinates are the same in the two media. In that case, $\boldsymbol{k}_s$ vectors in the two PMLs are identical, in place of (3.35), and $\boldsymbol{k}_{s1}$ replaces $\boldsymbol{k}_1$ in (3.37). This yields a set of four Eqs. (3.18) like at an interface between two identical PML media.

The time domain equations corresponding to the frequency domain PML (3.34) are derived in Chapter 4 for lossy, anisotropic, and dispersive media.

3.6   THE PML MATCHED TO NONHOMOGENEOUS MEDIA

In some applications of numerical methods, a PML is needed on the boundary of a computational domain composed of several media. In this paragraph, we consider the case where a PML is set perpendicularly to the interface between two physical media, as shown in Fig. 3.2. For example, this is the case in computational domains that extend both in the air and in the ground.

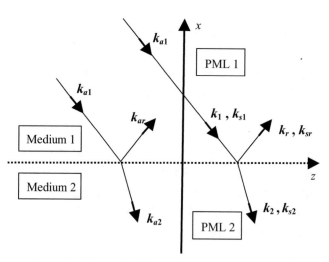

FIGURE 3.2: PML for nonhomogeneous media

We assume that two general anisotropic media are separated with an interface normal to x, and that two PMLs perpendicular to z have been obtained by stretching the z coordinate with the same factor s_z. Let us consider an incident wave in medium 1 with wave vector $\boldsymbol{k}_{a1}$. Relationship (3.37) holds for $\boldsymbol{k}_{a1}$ and the $\boldsymbol{k}_s$ vector of the wave transmitted into PML 1, with only changes in notations. Here

$$\overrightarrow{k_{s1}} = \overrightarrow{k_{a1}}. \tag{3.39}$$

At the interface between PML 1 and PML 2, we have $k_{1y} = k_{ry} = k_{2y}$ and $k_{1z} = k_{rz} = k_{2z}$. From this $k_{s1y} = k_{sry} = k_{s2y}$, and since the stretch of z is the same in the two PMLs, $k_{s1z} = k_{srz} = k_{s2z}$. Finally, using (3.39)

$$k_{s2y} = k_{sry} = k_{a1y} \quad k_{s2z} = k_{srz} = k_{a1z}. \tag{3.40}$$

Consider now the wave transmitted in medium 2, with wave vector $\boldsymbol{k}_{a2}$. Equality of the components of the $\boldsymbol{k}$ vectors in the interface between the two media yields

$$k_{a2y} = k_{ary} = k_{a1y} \quad k_{a2z} = k_{arz} = k_{a1z}. \tag{3.41}$$

Thus, from (3.40) and (3.41) two components of $\boldsymbol{k}_{a2}$ and $\boldsymbol{k}_{s2}$ are equal, so that the third components are also equal from (3.35). This holds for the reflected wave as well, and finally we can write

$$\overrightarrow{k_{sr}} = \overrightarrow{k_{ar}} \quad \overrightarrow{k_{s2}} = \overrightarrow{k_{a2}}. \tag{3.42}$$

Consider now the continuity of transverse components at the interface between PML 1 and PML 2. For the $\boldsymbol{E}$ polarization perpendicular to x direction, the set (3.18) holds with z coordinate in place of x because the interface is perpendicular to x instead of z. Using (3.34), Eq. (3.18d) becomes

$$\left[\overline{\overline{\mu_1}}^{-1} \overrightarrow{k_{s1}} \times \overrightarrow{E_{01}} \right]_y + \left[\overline{\overline{\mu_1}}^{-1} \overrightarrow{k_{sr}} \times \overrightarrow{E_{0r}} \right]_y = \left[\overline{\overline{\mu_2}}^{-1} \overrightarrow{k_{s2}} \times \overrightarrow{E_{02}} \right]_y \tag{3.43}$$

that can be rewritten with (3.39) and (3.42):

$$\left[\overline{\overline{\mu_1}}^{-1} \overrightarrow{k_{a1}} \times \overrightarrow{E_{01}} \right]_y + \left[\overline{\overline{\mu_1}}^{-1} \overrightarrow{k_{ar}} \times \overrightarrow{E_{0r}} \right]_y = \left[\overline{\overline{\mu_2}}^{-1} \overrightarrow{k_{a2}} \times \overrightarrow{E_{02}} \right]_y . \tag{3.44}$$

This is the equation that would be obtained at the interface between the nonstretched media 1 and 2. This also holds for the z component of (3.18). Finally (3.18) yields a set of equations identical to the set corresponding to the interface between the two nonstretched media. The conclusion is that the reflection coefficient between the two PMLs equals the reflection coefficient between the two media. This also holds for the transmission coefficient, and obviously for the other polarization of the incident wave as well.

In summary, in a nonhomogeneous computational domain like the one in Fig. 3.2, continuity of the reflection and transmission coefficients between media of the inner domain is ensured within the PML. This means that the reflected and transmitted waves are continuous at the inner domain–PML interface. No spurious field is produced. The simulation of free space is perfect in the presence of the media interface, provided that the two PMLs are stretched with the same factor.

3.7   THE UNIAXIAL PML MEDIUM

With the split PML (3.1) the equations governing the medium differ from the Maxwell equations, the PML is termed as nonMaxwellian in the literature. This renders its implementation difficult in such numerical methods as the finite element method. To overcome this problem, other formulations were developed like the one reported in Section 3.4. But the most popular unsplit PML is the uniaxial PML initially presented in [20] for use with the finite element method (FEM). This is an anisotropic Maxwellian medium where the field differs from that in the initial PML by a scale factor, but whose nonreflecting absorption is preserved.

The uniaxial PML can be derived from the PML (3.5) by means of the following change of variables:

$$E'_{0x} = s_x\,E_{0x} \quad E'_{0y} = s_y\,E_{0y} \quad E'_{0z} = s_z\,E_{0z} \tag{3.45a}$$

$$H'_{0x} = s^*_x\,H_{0x} \quad H'_{0y} = s^*_y\,H_{0y} \quad H'_{0z} = s^*_z\,H_{0z}. \tag{3.45b}$$

With this, Eq. (3.5a) becomes

$$\omega\varepsilon_0\frac{1}{s_x}E'_{0x} = -\frac{1}{s_y\,s^*_z}k_y\,H'_{0z} + \frac{1}{s_z\,s^*_y}k_z\,H'_{0y}. \tag{3.46}$$

If the matching condition holds, $s_x = s^*_x$, $s_y = s^*_y$, $s_z = s^*_z$, so that (3.46) can be rewritten as

$$\omega\varepsilon_0\frac{s_y\,s_z}{s_x}E'_{0x} = -k_y\,H'_{0z} + k_z\,H'_{0y}. \tag{3.47}$$

Similar equations are found from the other five equations of (3.5). The resulting system can be rewritten as:

$$\omega\varepsilon_0\overline{\overline{\varepsilon_s}}\,\vec{E'_0} = -\vec{k} \times \vec{H'_0} \tag{3.48a}$$

$$\omega\mu_0\overline{\overline{\mu_s}}\,\vec{H'_0} = \vec{k} \times \vec{E'_0} \tag{3.48b}$$

where tensors $\overline{\overline{\varepsilon_s}}$ and $\overline{\overline{\mu_s}}$ read:

$$\overline{\overline{\varepsilon_s}} = \overline{\overline{\mu_s}} = \begin{bmatrix} \dfrac{s_y s_z}{s_x} & 0 & 0 \\ 0 & \dfrac{s_z s_x}{s_y} & 0 \\ 0 & 0 & \dfrac{s_x s_y}{s_z} \end{bmatrix}. \qquad (3.49)$$

Equations (3.48) are the frequency domain Maxwell equations of an anisotropic medium of permittivity and permeability (3.49), called the uniaxial PML medium. Equations (3.48) can be used in a natural way with the frequency domain finite element method [21]. Time domain equations will be derived in the next chapter.

The absence of reflection from an interface located between two split PMLs whose transverse conductivities are equal is preserved with the uniaxial PML. In a computational domain surrounded with a PML ABC, the fields computed with the split PML (3.1) and with the uniaxial PML are identical in the inner domain, while in the PMLs they differ with ratios equal to the scaling factors in (3.45). The split PML is a stretch of coordinates while the uniaxial PML is a stretch of fields.

3.8 THE COMPLEX FREQUENCY SHIFTED PML

In the previous paragraphs, the stretching factors always are assumed of the form (3.3). In fact, the absence of reflection from a vacuum–PML interface, or more generally from PML–PML interfaces, also holds with any stretching factor. This is because the derivations in (3.15)–(3.25), and later with more general media, are valid with any s_x, s_y, s_z. What is required is only that the transverse stretches are equal on the two sides of the interface, and that the electric and magnetic stretches are equal, i.e., $s_u = s_u^*$ for $u = x, y, z$. From this, any other stretching factor can be envisaged in PMLs. Obviously, a requirement is that the stretching factor renders the PML lossy.

The simplest stretching factor is a real constant. Such a pure real stretch of the coordinate perpendicular to the interface produces no absorption. It has been used implicitly in the past in FDTD computer codes where the cells were stretched so as to move away the ABC while letting unchanged the overall number of cells (see the discussion on the location of traditional ABCs in Chapter 1). Use of this stretch is limited for numerical reasons because it rapidly degrades the sampling of the shortest wavelengths. A real stretch was also introduced in the context of the PML concept either to increase the natural decrease of evanescent waves [22], or to reduce the conductivity in the PML [23]. This simply consists in replacing unity in (3.3) with a real constant larger than unity. Nevertheless, use of a real stretch in a PML is also limited in magnitude because it increases the numerical reflection of traveling waves from the PML, due to the coarser sampling of shortest wavelengths. As shown in the following, a better solution to deal with evanescent waves is provided with the complex frequency shifted (CFS) PML.

The CFS PML was introduced by Kuzuoglu and Mittra [24] with the intention of rendering the PML medium causal. They added a degree of freedom α_u to the stretching factor. With in addition a real stretch κ_u in place of unity in (3.3), the CFS PML factors are defined as

$$s_u = \kappa_u + \frac{\sigma_u}{\alpha_u + j\omega\varepsilon_0} \quad (u = x, y, z).$$

(3.50)

Notice that α_u is homogeneous to a conductivity. In the following we assume that the PML medium is a wall PML perpendicular to x direction, only stretched in x direction. It is obvious from (3.50) that the features of the CFS-PML medium depend on the ratio of the frequency of interest to the frequency

$$f_\alpha = \frac{\alpha_x}{2\pi\varepsilon_0}.$$

(3.51)

For $f \gg f_\alpha$, parameter α_x is negligible in (3.50), so that the medium is a regular PML. Conversely, for $f \ll f_\alpha$ the stretch of x coordinate is real and the medium no longer absorbs the waves.

Due to the absence of absorption at low frequency, the CFS PML could appear as of little interest for use as an ABC in numerical methods. In fact, it is very well suited to the optimum absorption of frequency spectra composed of both evanescent and traveling waves present in many realistic physical problems. As derived in Chapter 2, the theoretical absorption of evanescent waves within a regular PML (2.35) may be so enormous that it cannot be properly sampled in space by finite methods. This results in a strong or even total numerical reflection from the PML ABC. As shown in the following, the CFS PML based on the stretching factor (3.50) allows the enormous absorption of evanescent waves to be reduced to a more reasonable value that can be sampled by numerical methods without significant spurious reflection [25].

We are now mainly interested in the waveform within the CFS PML. We consider the 2D case that has been generalized to evanescent waves in Chapter 2. In the CFS PML the wave numbers (2.27) also hold, with s_u factors (3.50) instead of (2.6). By inserting these wave numbers into the sinusoidal waveform (2.4), the counterpart of (2.30) is obtained for a matched CFS-PML perpendicular to x ($\sigma_y = 0$). If in addition the (X, Y) coordinates defined in Fig. 2.6 are used, the CFS-PML waveform can be expressed as follows:

$$\psi = \psi_0\, e^{j\omega\left[t - \frac{\cosh\chi}{c} X\right]}\, e^{-\frac{\omega}{c}\sinh\chi\, Y}\, e^{-j\frac{\omega}{c}\left[\kappa_x - 1 + \frac{\sigma_x}{j\omega\varepsilon_0 + \alpha_x}\right]C(\chi,\theta)x}.$$

(3.52)

The first two exponentials are nothing but the waveform of a nonhomogeneous wave propagating in X direction and evanescent in Y direction in a vacuum. The additional exponential accounts for the CFS-PML. As expected, at high frequency, i.e., if $f \gg f_\alpha$, with $\kappa_x = 1$ the waveform (3.52) reduces to the waveform in the split PML (2.31). Conversely, if $f \ll f_\alpha$, (3.52) can be rewritten as

$$\psi = \psi_{\text{vacuum}}\, e^{-j\frac{\omega}{c}\left(\kappa_x - 1 + \frac{\sigma_x}{\alpha_x}\right)\cosh\chi\,\cos\theta x}\, e^{\frac{\omega}{c}\left(\kappa_x - 1 + \frac{\sigma_x}{\alpha_x}\right)\sinh\chi\,\sin\theta x} \qquad (3.53)$$

where ψ_{vacuum} is the product of the first two exponentials in (3.52). The first exponential in (3.53) is an additional phase term. The second exponential is an absorbing term. Its argument is negative for waves propagating toward $+x$ and evanescent toward $+x$, because the signs of θ and χ are opposite in that case (either $\theta > 0$ and $\chi < 0$, or $\theta < 0$ and $\chi > 0$). The attenuation rate mainly depends on $\sigma_x \sinh\chi/\alpha_x$. In the case where $\kappa_x = 1$, using (3.51) the magnitude of the wave can be rewritten as

$$|\psi_{\text{PML}}| = |\psi_{\text{vacuum}}|\, e^{\frac{f}{f_\alpha}\frac{\sigma_x}{\varepsilon_0 c}\sinh\chi\,\sin\theta x}. \qquad (3.54)$$

Notice that the behavior of the wave in x direction could also be obtained by inserting the wave number k_x (2.27a) into $\exp(-j\,k_x x)$. For $f \ll f_\alpha$ this yields the following attenuation coefficient:

$$C_x = e^{\frac{\omega}{c}\left(\kappa_x + \frac{\sigma_x}{\alpha_x}\right)\sinh\chi\,\sin\theta x}. \qquad (3.55)$$

Coefficient (3.55) gives the overall decrease of the wave in x direction, while the coefficients in (3.53) and (3.54) only give the extra decrease due to the CFS-PML. The natural decrease is included in ψ_{vacuum}.

Since $\sinh\chi = 0$ for traveling waves, only evanescent waves are attenuated below f_α. Strictly speaking, coefficients (3.54) and (3.55) are not absorbing coefficients. It is evident from (3.55) that they only account for the natural decrease of the evanescent wave upon the stretched length $(\kappa_x + \sigma_x/\alpha_x)\, x$. Whatever the interpretation of these coefficients may be, what is important is that the extra attenuation in (3.54) is far smaller than its counterpart in a normal PML (2.35) for strongly evanescent waves and $f \ll f_\alpha$. This is because $\sinh\chi \approx \cosh\chi$ for $\chi \gg 1$ so that coefficients in (3.54) and (2.35) are about equal for $f = f_\alpha$, and then coefficient in (3.54) is far closer to unity for $f \ll f_\alpha$. From this, and by means of an adequate choice of the transition frequency (3.51), the CFS PML can allow the attenuation to be reasonable, i.e., neither too strong nor too weak, at both the evanescent and traveling frequencies present in actual physical problems. This is illustrated in the following with a waveguide problem.

Let us consider a parallel-plate waveguide where the modes are evanescent (1.4) below the cutoff frequency. Notice that the waveform (1.4a) is consistent with the general waveform in a vacuum, given by (2.30) with $\sigma_x = 0$. For a wave evanescent toward $+x$, (1.4a) corresponds to $\theta = \pi/2$ and $\chi < 0$, or $\theta = -\pi/2$ and $\chi > 0$, in (2.30). Far below cutoff (1.4c), the following

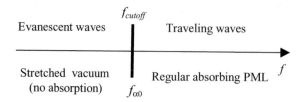

FIGURE 3.3: Coincidence of the transition frequency of the CFS-PML with the cutoff frequency of the parallel-plate waveguide

holds:

$$\omega \sinh \chi = \pm \frac{n\pi c}{a} \quad \text{for} \quad \omega \ll \frac{n\pi c}{a} \tag{3.56}$$

so that the CFS-PML attenuation in (3.54) no longer depends on frequency. By an adequate choice of the free parameter α_x this attenuation can be set equal to that of traveling waves at $f > f_\alpha$ and normal incidence. This is realized by equating the argument of the exponential in (3.54) with the argument of $\exp(-\sigma_x x/\varepsilon_0 c)$. Using (3.51) and since $\theta = \pm \pi/2$ and $\theta \chi < 0$ in the guide, this yields

$$-\frac{\sigma_x}{\varepsilon_0 c} = -\frac{2\pi \varepsilon_0}{\alpha_x} \frac{\sigma_x}{\varepsilon_0 c} \frac{1}{2\pi} \frac{n\pi c}{a}. \tag{3.57}$$

This equation gives α_x that renders the attenuation of low frequency evanescent waves equal to that of high frequency traveling waves:

$$\alpha_0 = n \frac{\pi c \varepsilon_0}{a}. \tag{3.58}$$

From (3.51), (3.58), and (1.4c), frequency f_α corresponding to α_0 is then

$$f_{\alpha 0} = \frac{nc}{2a} = f_{\text{cutoff}}. \tag{3.59}$$

Thus, with $\alpha_x = \alpha_0$ (3.58) the frequency f_α of the transition between the two regimes of the CFS PML is also the frequency f_{cutoff} of the transition between the evanescent and traveling waves of the considered mode in the waveguide. This is summarized in Fig. 3.3.

In summary, with α_x (3.58) the absorption is about uniform at all the frequencies, with the exception of a narrow band centered about the cutoff frequency where the attenuation vanishes because at the cutoff the propagation is parallel to the vacuum–PML interface. As will be shown in Chapter 5, a uniform attenuation is preferable in numerical methods, because an enormous attenuation as that produced by the regular PML (2.35) results in a strong numerical reflection. The CFS PML is very well suited to the considered waveguide problem. This also holds in other physical problems where evanescent waves are present at low frequency with the product $\omega \sinh\chi$ about constant like in (3.56), allowing then the attenuation to be about uniform and reasonable at all the frequencies of interest. This is the case in wave-structure interaction problems and in problems involving Floquet modes. Both problems are addressed in Chapter 6.

CHAPTER 4

Time Domain Equations
for the PML Medium

In Chapter 3, the PML medium matched to a vacuum has been introduced in time domain (3.1), from which the frequency domain equations (3.5) were derived. The PML (3.1) is known as the split PML. In the first part of the present chapter, it is shown that other time domain equations can be derived from (3.5). These time domain PMLs are the convolutional PML (CPML) and the near PML (NPML). The field components are not split within the CPML and the NPML, but auxiliary variables are needed to account for the PML medium, as with the dependent current interpretation addressed in Section 3.4. In this part, time domain equations are also provided for the uniaxial PML (3.48).

The last three parts of this chapter are devoted to the derivation of time domain counterparts of the frequency domain equations of the PML matched to general media (3.34). In the second part, time domain equations are derived for the split PML, the CPML, the NPML, and the uniaxial PML, in the case of a PML matched to isotropic lossy media. In the third part, time domain equations are provided for PMLs matched to anisotropic media. And in the fourth part, PMLs matched to dispersive media are addressed.

4.1 TIME DOMAIN PML MATCHED TO A VACUUM

In the following, three sets of time domain equations are derived from the frequency domain equations (3.5), namely the split PML, the CPML, and the NPML. Equations for the uniaxial PML are derived from (3.48).

4.1.1 The Split PML

Since the introduction of the PML concept [5], the time domain equations of the split PML (3.1) have been widely used in computational electromagnetics. Nevertheless, in a recent paper [26], it was shown that the requirements of the split PML can be reduced to only eight variables to be stored in the walls of the PML, in place of ten variables with (3.1). These new equations are derived in the following. To this end, let us consider a PML perpendicular to x, where

$\sigma_y = \sigma_z = 0$ so that the loss term is only present in Eqs. (3.1d), (3.1e) and (3.1j), (3.1k). Let us define two auxiliary variables as

$$\chi_{ex}(t) = \frac{1}{\mu_0} \int_{-\infty}^{t} E_x(t')dt' \qquad (4.1a)$$

$$\chi_{bx}(t) = \frac{1}{\varepsilon_0} \int_{-\infty}^{t} H_x(t')dt' \qquad (4.1b)$$

where $E_x = E_{xy} + E_{xz}$ and $H_x = H_{xy} + H_{xz}$. By integrating on time equations (3.1c), (3.1f), (3.1i), (3.1l), we can write

$$E_{yz} = \frac{\partial \chi_{bx}}{\partial z} \qquad E_{zy} = -\frac{\partial \chi_{bx}}{\partial y} \qquad (4.2)$$

$$H_{yz} = -\frac{\partial \chi_{ex}}{\partial z} \qquad H_{zy} = \frac{\partial \chi_{ex}}{\partial y}. \qquad (4.3)$$

This shows that subcomponents E_{yz} and E_{zy} can be obtained at any time from a unique quantity χ_{bx}. Similarly, H_{yz} and H_{zy} can be obtained from χ_{ex}. From this, the four Eqs. (3.1c)–(3.1f) can be replaced with only three equations, namely (4.1b) and two modified Eqs. (3.1d) and (3.1e):

$$\varepsilon_0 \frac{\partial E_{yx}}{\partial t} + \sigma_x E_{yx} = -\frac{\partial H_{zx}}{\partial x} - \frac{\partial^2 \chi_{ex}}{\partial x \partial y} \qquad (4.4a)$$

$$\varepsilon_0 \frac{\partial E_{zx}}{\partial t} + \sigma_x E_{zx} = \frac{\partial H_{yx}}{\partial x} - \frac{\partial^2 \chi_{ex}}{\partial x \partial z}. \qquad (4.4b)$$

Similarly (3.1i)–(3.1l) are replaced with (4.1a) and two dual equations of (4.4). Finally, system (3.1) reduces to eight equations involving eight variables to be stored, E_x, χ_{ex}, E_{yx}, E_{zx}, H_x, χ_{bx}, H_{yx}, H_{zx}. In the edge and corner regions, the number of variables cannot be reduced so that (3.1) must be used in numerical methods.

4.1.2   The Convolutional PML

The convolutional PML (CPML) is a set of time domain equations [27] derived from the frequency domain equations (3.5). The major interest of the CPML equations is its easy generalization to any physical medium. Consider for instance (3.5a), or equivalently (3.27a). In time domain the latter can be written as:

$$\varepsilon_0 \frac{\partial E_x}{\partial t} = \bar{s}_y(t) * \frac{\partial H_z}{\partial y} - \bar{s}_z(t) * \frac{\partial H_y}{\partial z} \qquad (4.5)$$

where $*$ is the convolution product, and $\bar{s}_y(t)$ and $\bar{s}_z(t)$ are the inverse Laplace transforms of the inverses of the stretching factors $s_y(\omega)$ and $s_z(\omega)$. Consider now stretching factors of the form (3.3). The inverse Laplace transform of the inverse of (3.3) reads:

$$\overline{s_u}(t) = \delta(t) + \zeta_u(t) \tag{4.6a}$$

where $\delta(t)$ is the Dirac function, and

$$\zeta_u(t) = -\frac{\sigma_u}{\varepsilon_0} e^{-\frac{\sigma_u}{\varepsilon_0}t} u(t) \tag{4.6b}$$

where $u(t)$ is the unit step function. Then, Eq. (4.5) can be rewritten as

$$\varepsilon_0 \frac{\partial E_x}{\partial t} = \frac{\partial H_z}{\partial y} - \frac{\partial H_y}{\partial z} + \psi_{hzy} - \psi_{hyz} \tag{4.7}$$

where

$$\psi_{hzy} = \zeta_y(t) * \frac{\partial H_z}{\partial y} \tag{4.8a}$$

$$\psi_{hyz} = \zeta_z(t) * \frac{\partial H_y}{\partial z}. \tag{4.8b}$$

Five equations similar to (4.7) are obtained from (3.5b)–(3.5f) for the time derivatives of the other five components of the field. This system is equivalent to the Maxwell equations, with the addition of auxiliary quantities $\psi_{ezy}, \psi_{eyz}, \dots, \psi_{hxy}, \psi_{hyx}$. There are 12 additional quantities in the case of a PML medium stretched in the three dimensions. In the walls of a PML, where only one direction is stretched, there are only four ψ variables, for instance $\psi_{eyz}, \psi_{exz}, \psi_{hxz}, \psi_{hyz}$ in a PML perpendicular to z. In the walls, the storage requirements are then the same as with the split PML (3.1), while in the edge and corner regions 18 variables have to be stored and advanced, instead of 12 with (3.1).

In time domain methods, the advance of the auxiliary quantities requires a convolution. In general this is a large computational burden. In the case of exponential impulses like (4.6b) the convolution is recursive and can be performed with a small number of operations. This is illustrated with the FDTD method in Chapter 5.

In the case where the CFS stretching factor (3.50) is used, the inverse Laplace transform (4.6) is replaced with:

$$\overline{s_u}(t) = \frac{\delta(t)}{\kappa_u} + \zeta_u(t) \tag{4.9a}$$

$$\zeta_u(t) = -\frac{\sigma_u}{\varepsilon_0 \kappa_u^2} e^{-\frac{1}{\varepsilon_0}\left(\frac{\sigma_u}{\kappa_u}+\alpha_u\right)t} u(t) \tag{4.9b}$$

and Eq. (4.5) yields in place of (4.7)

$$\varepsilon_0 \frac{\partial E_x}{\partial t} = \frac{1}{\kappa_y} \frac{\partial H_z}{\partial y} - \frac{1}{\kappa_z} \frac{\partial H_y}{\partial z} + \psi_{hzy} - \psi_{hyz} \qquad (4.10)$$

where ψ_{hzy} and ψ_{hyz} are also given by (4.8).

4.1.3 The Near PML

The near PML (NPML) was introduced [28] with the intention of simplifying the implementation of the PML ABC in problems where the inner medium is the gyrotropical ionosphere. It was called a near PML because it is equivalent to a true PML only in the case where the conductivity does not depend on space coordinates. Nevertheless, numerical experiments in [28] showed no more reflection from the NPML ABC than from the regular PML ABC. The absence of reflection from vacuum–NPML or NPML–NPML interfaces was proved theoretically later [29, 30].

Let us start from Eqs. (3.5) or (3.27). Assuming that the conductivities and then the stretching factors are constant in the PML, (3.27a) can be rewritten as

$$j\omega\varepsilon_0 E_x = \frac{\partial\left(H_z/s_y\right)}{\partial y} - \frac{\partial\left(H_y/s_z\right)}{\partial z} \qquad (4.11)$$

or equivalently

$$j\omega\varepsilon_0 E_x = \frac{\partial\xi_{hzy}}{\partial y} - \frac{\partial\xi_{hyz}}{\partial z} \qquad (4.12a)$$

where

$$\xi_{hzy} = \frac{H_z}{s_y} \quad \xi_{hyz} = \frac{H_y}{s_z}. \qquad (4.12b)$$

Similar manipulations can be done with (3.5b)–(3.5f), resulting in the definition of 12 variables ξ_e and ξ_h in addition to the six components of the field. The time domain counterpart of (4.12a) reads

$$\varepsilon_0 \frac{\partial E_x}{\partial t} = \frac{\partial\xi_{hzy}}{\partial y} - \frac{\partial\xi_{hyz}}{\partial z} \qquad (4.13)$$

and by using (3.3) the time domain counterpart of (4.12b) can be written as

$$\frac{\partial\xi_{hzy}}{\partial t} + \frac{\sigma_y}{\varepsilon_0}\xi_{hzy} = \frac{\partial H_z}{\partial t} \qquad (4.14a)$$

$$\frac{\partial\xi_{hyz}}{\partial t} + \frac{\sigma_z}{\varepsilon_0}\xi_{hyz} = \frac{\partial H_y}{\partial t}. \qquad (4.14b)$$

Equation (4.13) and the other five equations obtained from (3.5b)–(3.5f) are identical to the Maxwell equations, with E and H components replaced with auxiliary quantities ξ_e and ξ_h in the curls. From this, in computational methods the algorithm used in the inner domain can also be used in the NMPL. In addition, two differential equations have to be solved per component, for instance (4.14) for E_x.

As assumed in the above as incorporating the stretching factors in the space derivatives in (4.11), the NPML is identical to the true PML only in the case where the stretching factors do not depend on space coordinates. Nevertheless, as proved in [29, 30], at the interface between two NPMLs having different conductivities, or between a vacuum and a NPML, there is no reflection. From this, for use as an ABC in numerical methods, the NPML is equivalent to a true PML, because in that case the conductivity is constant in the elementary volumes or cells, so that the actual PML ABC is a juxtaposition of different NPMLs of constant conductivity with no reflection from their interfaces.

As with the CPML, in the general case there are 18 variables in the NPML, instead of 12 with the split PML. Nevertheless, in the walls of the NPML only ten variables have to be computed and stored, as with the split PML. For instance in a wall NPML perpendicular to z, only ξ_{exz}, ξ_{eyz}, ξ_{hxz}, ξ_{hyz} differ from the field components, because $s_x = s_y = 1$. The NPML can be implemented easily in any physical medium. For example, if the permittivity ε_0 is replaced with a tensor in (4.13) and in the corresponding E_y and E_z equations, the algorithm used in the inner domain is left unchanged in the NPML. The only modification is the solution of 12 differential equations like (4.14) in the NPML (four equations in the walls).

4.1.4   The Uniaxial PML

Let us consider frequency domain equations (3.48) of the uniaxial PML. In the following the prime is omitted, the fields in the uniaxial PML are simply denoted as $E_x, \ldots, H_z$. The first component of (3.48a) can be rewritten as

$$j\omega\varepsilon_0 \frac{s_y s_z}{s_x} E_x = \frac{\partial H_z}{\partial y} - \frac{\partial H_y}{\partial z}. \qquad (4.15)$$

In time domain, a convolution is needed in the left-hand member. To remove it, an auxiliary variable D_x is introduced [31],

$$D_x = \varepsilon_0 \frac{s_z}{s_x} E_x, \qquad (4.16)$$

so that (4.15) becomes

$$j\omega s_y D_x = \frac{\partial H_z}{\partial y} - \frac{\partial H_y}{\partial z}. \qquad (4.17)$$

Replacing then s_x, s_y, s_z with their explicit values (3.3) and performing an inverse Fourier transform, that is replacing $j\omega$ with the time derivative, the time domain counterparts of (4.16) and (4.17) read

$$\varepsilon_0 \frac{\partial E_x}{\partial t} + \sigma_z E_x = \frac{\partial D_x}{\partial t} + \frac{\sigma_x}{\varepsilon_0} D_x \qquad (4.18)$$

$$\frac{\partial D_x}{\partial t} + \frac{\sigma_y}{\varepsilon_0} D_x = \frac{\partial H_z}{\partial y} - \frac{\partial H_y}{\partial z}. \qquad (4.19)$$

Five similar sets of two equations are derived from the remaining five components of (3.48). Equation (4.19) is like a component of the Maxwell–Ampere equation in a lossy medium, and (4.18) is a differential equation, so that the uniaxial PML can be regarded as a lossy medium with a set of six differential equations.

The application of the uniaxial PML to the FDTD method is described in [31]. In some papers, the uniaxial PML is denoted as unsplit PML, although other unsplit PMLs do exist, namely the CPML and the NPML. But the uniaxial PML has the unique feature of being Maxwellian, that is satisfying the Maxwell equations.

4.2   TIME DOMAIN PML FOR LOSSY MEDIA

In the following, the time domain equations of the PML matched to an isotropic lossy medium are derived. As in the case of a vacuum, four versions of the PML for lossy media are derived, namely the split PML, the CPML, the NPML, and the uniaxial PML.

4.2.1   Split PML for Lossy Media

Let us consider the frequency domain Maxwell equations of a lossy medium of conductivity σ. As shown in Chapter 3 for any general medium, the equations of the corresponding PML are obtained by stretching the coordinates. From this, the PML equations for lossy media read

$$\omega\varepsilon_0 \left(1 + \frac{\sigma}{j\omega\varepsilon_0}\right) \vec{E_0} = -\vec{k_s} \times \vec{H_0} \qquad (4.20a)$$

$$\omega\mu_0 \vec{H_0} = \vec{k_s} \times \vec{E_0}. \qquad (4.20b)$$

The x component of (4.20a) can be written as

$$\omega\varepsilon_0 \left(1 + \frac{\sigma}{j\omega\varepsilon_0}\right) E_{0x} = -\frac{k_y}{s_y} H_{0z} + \frac{k_z}{s_z} H_{0y}. \qquad (4.21)$$

As in a vacuum, (4.21) can be split into two subequations with E_{0x} component split into two subcomponents E_{0xy} and E_{0xz}:

$$\omega\varepsilon_0 \left(1 + \frac{\sigma}{j\omega\varepsilon_0}\right) E_{0xy} = -\frac{k_y}{s_y} H_{0z} \tag{4.22a}$$

$$\omega\varepsilon_0 \left(1 + \frac{\sigma}{j\omega\varepsilon_0}\right) E_{0xz} = \frac{k_z}{s_z} H_{0y}. \tag{4.22b}$$

By replacing the stretching factor s_y with (3.3), Eq. (4.22a) becomes

$$\left(j\omega\varepsilon_0 + \sigma + \sigma_y + \frac{\sigma\sigma_y}{j\omega\varepsilon_0}\right) E_{0xy} = -jk_y H_{0z}. \tag{4.23}$$

Since dividing with $j\omega$ in frequency domain corresponds to integrating on time, the time domain counterpart of (4.23) is

$$\varepsilon_0 \frac{\partial E_{xy}}{\partial t} + (\sigma + \sigma_y)E_{xy} + \frac{\sigma\sigma_y}{\varepsilon_0} \int_{-\infty}^{t} E_{xy} dt' = \frac{\partial H_z}{\partial y}. \tag{4.24}$$

From (4.22b) a similar equation gives the time derivative of subcomponent E_{xz}. The y and z components of (4.20a) can also be split into subequations, with E_y and E_z split into subcomponents, resulting in four equations similar to (4.24). Finally, since Eq. (4.20b) is like in a PML matched to a vacuum, the time domain equations in the split PML matched to a lossy medium are those in the PML matched to a vacuum (3.1), with only the modification of the first six Eqs. (3.1a)–(3.1f) according to (4.24). The modification consists of the addition of the conductivity of the medium σ to the PML conductivity, and the addition of an extra term proportional to the medium conductivity, the PML conductivity, and the integral on time of the corresponding electric subcomponent.

As shown in the next chapter, the six equations like (4.24) can be discretized easily with the FDTD method. The above PML for lossy media was first presented in [32]. This was probably the first extension of the PML absorbing boundary to a medium that differs from a vacuum or a dielectric.

4.2.2   CPML for Lossy Media

Equation (4.21) in the PML matched to lossy media can be rewritten in time domain as:

$$\varepsilon_0 \frac{\partial E_x}{\partial t} + \sigma E_x = \overline{s_y}(t) * \frac{\partial H_z}{\partial y} - \overline{s_z}(t) * \frac{\partial H_y}{\partial z} \tag{4.25}$$

that is similar to (4.5), with only the addition of the term σE_x on the left-hand side of the equation. From this, the convolutions and the corresponding ψ variables defined in the case of the PML matched to a vacuum are left unchanged in the presence of a loss term in the medium.

This yields the following counterpart of (4.10):

$$\varepsilon_0 \frac{\partial E_x}{\partial t} + \sigma E_x = \frac{1}{\kappa_y} \frac{\partial H_z}{\partial y} - \frac{1}{\kappa_z} \frac{\partial H_y}{\partial z} + \psi_{hzy} - \psi_{hyz} \qquad (4.26)$$

where ψ quantities are also given by (4.8). Equation (4.26) is like that in the lossy medium, with the addition of the ψ quantities to the curl. Two similar equations are obtained for the evolution on time of E_y and E_z components. The three equations governing the components of the H field are left unchanged in comparison with the CPML matched to a vacuum. The implementation of the CPML in marching on time finite methods is easy. The discretized equations are like in the corresponding lossy medium, with only the trivial addition of ψ variables governed by equations like (4.8).

4.2.3   NPML for Lossy Media

In that case, with the variable changes (4.12b) Eq. (4.21) becomes in time domain:

$$\varepsilon_0 \frac{\partial E_x}{\partial t} + \sigma E_x = \frac{\partial \xi_{hzy}}{\partial y} - \frac{\partial \xi_{hyz}}{\partial z} \qquad (4.27)$$

where ξ_{hzy} and ξ_{hyz} are governed by (4.14) as well. Two similar equations hold for the evolution in time of E_y and E_z components. Finally, the six equations governing the field components in the PML are identical to the Maxwell equations of the lossy medium, with ξ variables instead of E or H components in the curls. From this, the implementation of a NPML matched to a lossy medium is trivial in finite methods. Only the advance of the ξ_e and ξ_h quantities by means of differential equations like (4.14), that are independent of the medium, has to be added to the algorithm.

4.2.4   Uniaxial PML for Lossy Media

A uniaxial PML for isotropic lossy media was presented in [33]. In the frequency domain it can be obtained from the PML matched to a vacuum by replacing ε_0 with the complex permittivity of the medium. From this, Eq. (3.48a) is replaced with

$$\omega \varepsilon_0 \left(1 + \frac{\sigma}{j\omega\varepsilon_0}\right) \overline{\overline{\varepsilon_s}} \vec{E_0} = -\vec{k} \times \vec{H_0} \qquad (4.28)$$

whose first component can be written as

$$j\omega\varepsilon_0 \left(1 + \frac{\sigma}{j\omega\varepsilon_0}\right) \frac{s_y s_z}{s_x} E_x = \frac{\partial H_z}{\partial y} - \frac{\partial H_y}{\partial z}. \qquad (4.29)$$

Let us now define two auxiliary variables as

$$D'_x = s_y D_x \tag{4.30a}$$

$$D_x = \varepsilon_0 \frac{s_z}{s_x} E_x. \tag{4.30b}$$

Then (4.29) can be rewritten as

$$j\omega \left(1 + \frac{\sigma}{j\omega\varepsilon_0}\right) D'_x = \frac{\partial H_z}{\partial y} - \frac{\partial H_y}{\partial z} \tag{4.31}$$

or in time domain:

$$\frac{\partial D'_x}{\partial t} + \frac{\sigma}{\varepsilon_0} D'_x = \frac{\partial H_z}{\partial y} - \frac{\partial H_y}{\partial z}. \tag{4.32}$$

Moreover, (4.30) yields in time domain:

$$\frac{\partial D_x}{\partial t} + \frac{\sigma_y}{\varepsilon_0} D_x = \frac{\partial D'_x}{\partial t} \tag{4.33a}$$

$$\varepsilon_0 \frac{\partial E_x}{\partial t} + \sigma_z E_x = \frac{\partial D_x}{\partial t} + \frac{\sigma_x}{\varepsilon_0} D_x. \tag{4.33b}$$

Equations (4.32) and (4.33b) are like (4.19) and (4.18) in the uniaxial PML matched to a vacuum, and (4.33a) is a special case of (4.33b). Then, these three equations can be discretized easily with finite methods. In marching on time methods, (4.32) is used to advance on time D'_x, then (4.33a) is used to advance D_x, and finally (4.33b) permits E_x to be advanced. Two sets similar to (4.32), (4.33) are obtained from (4.28) for the advance of components E_y and E_z. And since (3.48b) also holds in the lossy uniaxial PML, the equations for the advance of the magnetic field are like in the uniaxial PML matched to a vacuum.

4.3   TIME DOMAIN PML FOR ANISOTROPIC MEDIA

To derive time domain equations from the frequency domain equations (3.34) of a PML matched to any general medium two methods can be used:

1. incorporating the stretching coefficients on the left-hand side of Eqs. (3.34).
2. incorporating the stretching coefficients on the right-hand side of Eqs. (3.34).

With the isotropic lossy medium considered in the previous paragraph, method 1 corresponds to the split PML, while method 2 corresponds to the CPML and the NPML. With such a simple medium, the latter method is trivial and yields time domain equations close to the equations of the nonstretched lossy medium. That is also true when more general physical media are considered. Obtaining time domain equations is trivial and always possible by incorporating the stretching factors on the right-hand side of (3.34), whatever the medium may be.

4.3.1   Split PML for Anisotropic Media

Several papers were published in the literature on the generalization of the split PML to anisotropic media [34–39]. Two approaches are possible for nondispersive media. The first one consists of deriving equations and sub-equations involving only components and subcomponents of the E and H fields. The second one is based on use of fields D and B that are split in place of E and H. Discussions about the former approach can be found in [37]. The latter approach, which is more general and more easily derived, is presented in the following. To this end, the first component of (3.34) is rewritten as

$$\omega D_{0x} = -\frac{k_y}{s_y} H_{0z} + \frac{k_z}{s_z} H_{0y} \tag{4.34a}$$

where D_{0x} is the x component of the left-hand member of (3.34a):

$$D_{0x} = \left[\overline{\overline{\varepsilon}} \vec{E}_0 \right]_x . \tag{4.34b}$$

Equation (4.34a) and component D_{0x} can be split as

$$\omega D_{0xy} = -\frac{k_y}{s_y} H_{0z} \tag{4.35a}$$

$$\omega D_{0xz} = \frac{k_z}{s_z} H_{0y} . \tag{4.35b}$$

This yields the time domain equations that govern subcomponents D_{xy} and D_{xz}

$$\frac{\partial D_{xy}}{\partial t} + \frac{\sigma_y}{\varepsilon_0} D_{xy} = \frac{\partial H_z}{\partial y} \tag{4.36a}$$

$$\frac{\partial D_{xz}}{\partial t} + \frac{\sigma_z}{\varepsilon_0} D_{xz} = -\frac{\partial H_y}{\partial z} . \tag{4.36b}$$

Four similar equations are obtained from the other two components of (3.34a). With marching on time methods, this permits the six subcomponents of D to be advanced. Assuming that the medium is nondispersive, field E can then be computed by solving for E the algebraic equation:

$$\vec{D} = \overline{\overline{\varepsilon}} \vec{E} \tag{4.37}$$

where each D component is the sum of the corresponding two subcomponents. If only the permittivity is anisotropic in (3.34), (3.34b) is like in a vacuum and yields the same split equations (3.1g)–(3.1l). In the case where the permeability is a tensor, a B field is introduced in

(3.34b) and split into subcomponents as done in the above with $\boldsymbol{D}$. This yields dual equations of (4.36), (4.37).

4.3.2   CPML for Anisotropic Media

In time domain, the first component of frequency domain equation (3.34a) can be written as:

$$\frac{\partial \left[\overline{\overline{\varepsilon}}\,\vec{E}\right]_x}{\partial t} = \overline{s_y}(t) * \frac{\partial H_z}{\partial y} - \overline{s_z}(t) * \frac{\partial H_y}{\partial z} \qquad (4.38)$$

where it is assumed that tensor ε is not frequency dependent. The convolutions are identical to those in a PML matched to a vacuum (4.5), so that (4.38) can be rewritten as

$$\frac{\partial \left[\overline{\overline{\varepsilon}}\,\vec{E}\right]_x}{\partial t} = \frac{1}{\kappa_y}\frac{\partial H_z}{\partial y} - \frac{1}{\kappa_z}\frac{\partial H_y}{\partial z} + \psi_{hzy} - \psi_{hyz} \qquad (4.39)$$

where the ψ parameters again are given by (4.8). This equation is nothing but the corresponding equation in the anisotropic nonstretched medium, with the addition of ψ variables. Five similar equations are obtained from (3.34). Finally, the time domain equations of the PML are those of the corresponding anisotropic medium with only the addition of ψ variables to the curls, as in (4.39). The ψ variables are governed by the same Eqs. (4.8) as in a CPML matched to a vacuum. From this, the implementation of the CPML for anisotropic media in marching on time finite methods is straightforward.

4.3.3   NPML for Anisotropic Media

With the variable changes (4.12b), the x component of Eq. (3.34a) becomes in time domain:

$$\frac{\partial \left[\overline{\overline{\varepsilon}}\,\vec{E}\right]_x}{\partial t} = \frac{\partial \xi_{hzy}}{\partial y} - \frac{\partial \xi_{hyz}}{\partial z} \qquad (4.40)$$

where ξ_{hzy} and ξ_{hyz} are governed by (4.14) as well. Five similar equations hold for the evolution in time of E_y, E_z, H_x, H_y, and H_z components. Finally, the six equations governing the field components in the PML are identical to the Maxwell equations of the anisotropic medium, with ξ_e and ξ_h variables instead of $\boldsymbol{E}$ or $\boldsymbol{H}$ components in the curls. That is true whatever the tensors ε and μ may be. From this, the implementation of the NPML matched to any anisotropic medium is trivial in finite methods. Only the advance of the ξ quantities with equations like (4.14) has to be added to the algorithm.

4.4 TIME DOMAIN PML FOR DISPERSIVE MEDIA

In some physical media, the permittivity or the permeability depends on frequency. These media may be isotropic or anisotropic. Two important classes of material dispersion are the Debye relaxation and the Lorentzian resonance where ε depends on frequency. Another example of dispersive medium is the gyrotropic, i.e., anisotropic, Ionosphere. In such media, use of the split PML would not be easy nor effective, because a convolution would be present in (4.37). In the following we limit our purpose to the CPML and NPML cases whose time domain equations derivation is straightforward. The isotropic uniaxial PML is also addressed.

4.4.1 Time Domain CPML and NPML for Isotropic or Anisotropic Dispersive Media

Obtaining time domain equations for the CPML and NPML is trivial. In the case of a CPML matched to a dispersive anisotropic medium, the first component of (3.34a) is like (4.38), with a convolution in the left-hand member due to the frequency dependence of the permittivity tensor:

$$\frac{\partial \left[\bar{\bar{\varepsilon}} * \vec{E} \right]_x}{\partial t} = \bar{s_y}(t) * \frac{\partial H_z}{\partial y} - \bar{s_z}(t) * \frac{\partial H_y}{\partial z}. \qquad (4.41)$$

The convolutions in the right-hand member are identical to those in the CPML matched to a vacuum (4.5), so that (4.41) can be rewritten as

$$\frac{\partial \left[\bar{\bar{\varepsilon}} * \vec{E} \right]_x}{\partial t} = \frac{1}{\kappa_y} \frac{\partial H_z}{\partial y} - \frac{1}{\kappa_z} \frac{\partial H_y}{\partial z} + \psi_{hzy} - \psi_{hyz} \qquad (4.42)$$

where ψ variables are given by (4.48). Five similar equations hold for the other five components of the field. Once again, the CPML equations are like the ones in the corresponding nonstretched medium, with the addition of ψ variables to the curls. If a numerical algorithm is available for the inner domain, implementing the corresponding PML is straightforward.

Similarly, in the NPML the first component of (3.34a) yields:

$$\frac{\partial \left[\bar{\bar{\varepsilon}} * \vec{E} \right]_x}{\partial t} = \frac{\partial \xi_{hzy}}{\partial y} - \frac{\partial \xi_{hyz}}{\partial z} \qquad (4.43)$$

so that the equations in the NPML are identical to those in the corresponding inner domain with ξ_e and ξ_h variables (4.14) in place of **E** or **H** components in the curls.

4.4.2 Time Domain Uniaxial PML for Isotropic Dispersive Media

Use of the uniaxial PML with the FDTD method was reported [33] in the case of isotropic dispersive media where $\varepsilon = \varepsilon_0\, \varepsilon_r(\omega)$. In that case, (4.15) is replaced with

$$j\omega\varepsilon_0\varepsilon_r(\omega)\frac{s_y\,s_z}{s_x}E_x = \frac{\partial H_z}{\partial y} - \frac{\partial H_y}{\partial z} \tag{4.44}$$

that is close to (4.29) in the lossy medium case. It can be solved by the variable changes

$$D'_x = \varepsilon_r(\omega)D_x \tag{4.45a}$$

$$D_x = \varepsilon_0\frac{s_z}{s_x}E_x \tag{4.45b}$$

so that (4.44) becomes

$$j\omega\left(1 + \frac{\sigma_y}{j\omega\varepsilon_0}\right)D'_x = \frac{\partial H_z}{\partial y} - \frac{\partial H_y}{\partial z}. \tag{4.46}$$

Equations (4.45b) and (4.46) are identical to (4.30b) and (4.31) and yield time domain equations (4.33b) and (4.32). Equation (4.45a) is like the auxiliary equation used in the corresponding nonstretched dispersive medium [33], so that its time domain counterpart is like in this medium.

CHAPTER 5

The PML ABC for the FDTD Method

The finite-difference time-domain method (FDTD) [40] is probably the most used numerical technique in electromagnetics. The basic FDTD scheme introduced by K. S. Yee in 1966 [2] is simple and versatile. It has been extended to most existing physical media, such as anisotropic, dispersive, chiral, and even nonlinear media. The original Yee scheme is a second-order approximation of the Maxwell equations, both in time and space.

More sophistical schemes have been developed over the years, mainly with the intention of reducing the numerical dispersion of the Yee scheme. Four order schemes, pseudospectral time-domain (PSTD) schemes, multi-resolution time-domain (MRTD) schemes, all are schemes with widely smaller numerical dispersion [40]. But they are more complex and less versatile, so that nowadays the basic Yee scheme remains the most popular in applications. In this chapter we only address the implementation of the PML ABC in the Yee scheme. PMLs for other FDTD schemes can be found in the literature. References are provided in Chapter 7.

5.1 FDTD SCHEMES FOR THE PML MATCHED TO A VACUUM

FDTD schemes for the split PML, the CPML, the NPML, and the uniaxial PML are derived in the following.

5.1.1 FDTD Scheme for the Split PML

In the three-dimensional PML, 12 subcomponents have to be advanced on time (3.1) in each FDTD cell. The regular FDTD grid [40] is left unchanged, simply two subcomponents are computed at the location of each component. For instance, E_{xy} and E_{xz} are computed at E_x nodes of the regular FDTD grid. Discretization of (3.1) is straightforward. As an example, for the advance of E_{xy} from time n to time $n + 1$ at grid node $(i + 1/2, j, k)$, Eq. (3.1a) yields, with usual FDTD notations [40]:

$$E_{xy}\big|_{i+1/2,j,k}^{n+1} = A_y\, E_{xy}\big|_{i+1/2,j,k}^n + \frac{\Delta t}{\varepsilon_0}\, B_y\, \frac{[H_{zx} + H_{zy}]_{i+1/2,j+1/2,k}^{n+1/2} - [H_{zx} + H_{zy}]_{i+1/2,j-1/2,k}^{n+1/2}}{\Delta y}$$

$$(5.1)$$

where coefficients A_y and B_y, evaluated at E_x node $(i + 1/2, j, k)$, are of the form

$$A_u(\sigma_u) = \frac{2\varepsilon_0 - \sigma_u \Delta t}{2\varepsilon_0 + \sigma_u \Delta t} \tag{5.2a}$$

$$B_u(\sigma_u) = \frac{2\varepsilon_0}{2\varepsilon_0 + \sigma_u \Delta t}, \tag{5.2b}$$

or alternatively with an exponential discretization, valid with any value of the conductivity σ_u:

$$A_u(\sigma_u) = e^{-\sigma_u \frac{\Delta t}{\varepsilon_0}} \tag{5.3a}$$

$$B_u(\sigma_u) = \frac{\varepsilon_0(1 - A_u)}{\sigma_u \Delta t}. \tag{5.3b}$$

Eleven equations similar to (5.1) are obtained for the advance of the other 11 subcomponents.

In the walls of the PML (Fig. 3.1), two conductivities equal zero. From this, two sets of two subequations merge into single equations identical to the equations in a vacuum. For example, in a wall PML perpendicular to x, subequations (3.1a)–(3.1b) and (3.1g)–(3.1h) merge into two equations where E_{xy} and E_{xz} are replaced with E_x, and H_{xy} and H_{xz} are replaced with H_x. From this, only ten variables have to be advanced in time and stored in the walls of the PML. In the edges of the domain, where one conductivity equals zero, the conductivities that govern the two subcomponents of each couple of subcomponents are different, so that the 12 subcomponents must be computed and stored separately. This also holds in the corners where three different conductivities are present in each cell. Finally, with the split PML (3.1), ten quantities have to be stored in the walls that represent most of the PML, and 12 quantities in the remaining parts.

For the advance of components in the vacuum–PML interface, equations like (5.1) have to be slightly modified because the space derivative in the direction perpendicular to the interface involves one node in the vacuum and one node in the PML. The right equations are obtained by considering that the vacuum is nothing but a special case of PML media where the subcomponents are merged. From this, the two subcomponents are replaced with the corresponding component at the node in the vacuum. For example, $H_{zx} + H_{zy}$ is replaced with H_z in one bracket of (5.1) at an interface perpendicular to y. Similarly, at the nodes in the vacuum located half a cell from the interface, a component involved in the space derivative of the regular FDTD equation is in the interface, that is in the PML. This component is replaced with the sum of the corresponding two subcomponents.

In the case where the regular split equations (3.1) are replaced in the walls with the eight-variable split PML derived in Section 4.1.1, Eqs. (4.1) and (4.4) have to be discretized for the advance of the χ variables and the remaining subcomponents of the field. This can be done easily. For example (4.1a) and (4.4a) yield, respectively:

$$\chi_{ex}|_{i+1/2,j,k}^{n+1/2} = \chi_{ex}|_{i+1/2,j,k}^{n-1/2} + \frac{\Delta t}{\mu_0} \, E_x|_{i+1/2,j,k}^{n} \tag{5.4}$$

$$E_{yx}|_{i,j+1/2,k}^{n+1} = A_x \, E_{yx}|_{i,j+1/2,k}^{n} - \frac{\Delta t}{\varepsilon_0} B_x \frac{\Delta H_{zx}}{\Delta x}\Big|_{i,j+1/2,k}^{n+1/2}$$

$$- \frac{\Delta t}{\varepsilon_0} \frac{B_x}{\Delta x \Delta y} \left[\chi_{ex}|_{i+1/2,j+1,k}^{n+1/2} - \chi_{ex}|_{i+1/2,j,k}^{n+1/2} - \chi_{ex}|_{i-1/2,j+1,k}^{n+1/2} + \chi_{ex}|_{i-1/2,j,k}^{n+1/2} \right] \tag{5.5}$$

where $\Delta H_{zx}/\Delta x$ is the discretized space derivative of H_{zx} in x direction, and A_x and B_x are given by (5.2) or (5.3).

5.1.2   FDTD Scheme for the Convolutional PML

The six convolutional PML equations like (4.5) involve convolutions of the space derivatives of the field components with $\zeta_u(t)$ functions given either by (4.6b) for the regular PML or by (4.9b) for the CFS PML. In both cases $\zeta_u(t)$ is an exponential function. This makes it possible to perform a recursive convolution. Consider the convolution $g(t)$ of an exponential $\exp(-a\,t)$ with a function $f(t)$:

$$g(t) = \int_0^t f(t')e^{-a(t-t')}dt'. \tag{5.6}$$

It can be shown easily that, provided that Δt is small enough in order that the variation of $f(t')$ is small during interval $[t,\, t+\Delta t]$, so as to approximate it with $f(t+\Delta t/2)$, the following holds:

$$g(t+\Delta t) = e^{-a\Delta t}g(t) + f(t+\Delta t/2)\frac{1-e^{-a\Delta t}}{a}. \tag{5.7}$$

This will permit the calculation of the ψ functions in (4.8) at FDTD time $n+1$ in function of their values at the previous time n.

Consider now Eq. (4.10). It can be discretized as

$$E_x^{n+1} = E_x^n + \frac{\Delta t}{\varepsilon_0} \left[\frac{1}{\kappa_y} \frac{\Delta H_z^{n+1/2}}{\Delta y} - \frac{1}{\kappa_z} \frac{\Delta H_y^{n+1/2}}{\Delta z} + \psi_{hzy}^{n+1/2} - \psi_{hyz}^{n+1/2} \right] \tag{5.8}$$

where the space index $(i+1/2, j, k)$, where all the variables are localized, is omitted. The advance in time of ψ functions can be performed using the recursive formula (5.7). In the CFS PML case, (4.8a), (4.9b), and (5.7) lead to, at node $(i+1/2, j, k)$:

$$\psi_{hzy}^{n+1/2} = p_y \psi_{hzy}^{n-1/2} + q_y \frac{\Delta H_z^{n+1/2}}{\Delta y} \tag{5.9}$$

where

$$p_u = e^{-\left(\frac{\sigma_u}{\kappa_u} + \alpha_u\right)\frac{\Delta t}{\varepsilon_0}} \tag{5.10a}$$

$$q_u = \frac{\sigma_u}{\kappa_u(\sigma_u + \kappa_u\alpha_u)}(p_u - 1). \tag{5.10b}$$

Five equations like (5.8) are obtained for the advance of the other five components of the field, and 11 equations like (5.9) for the advance of the other ψ variables. For the six ψ_e variables, p and q are also given by (5.10), provided that the matching condition holds. In summary, using six equations like (5.8) and 12 equations like (5.9), the CPML algorithm consists in computing the E components and ψ_e variables at times n, $n+1$, and the H components and ψ_h variables at times $n-1/2$, $n+1/2$. Notice that the H field in (5.9) is not centered in time. The average with the previous value could be used, but in actual FDTD calculations (5.9) yields quite satisfactory results.

5.1.3 FDTD Scheme for the NPML

The FDTD discretization of the NPML equations is straightforward. The six equations for the advance of the components of the field, like (4.13), are nothing but the Maxwell equations. They are discretized as Maxwell equations, with ξ_e or ξ_h variables on the right-hand side in place of E or H components. The ξ variables are advanced in time with differential equations (4.14) and five similar sets. For instance, (4.14a) yields for ξ_{hzy}

$$\xi_{hzy}^{n+1/2} = A_y\xi_{hzy}^{n-1/2} + B_y\left[H_z^{n+1/2} - H_z^{n-1/2}\right] \tag{5.11}$$

where A_y and B_y are given by (5.2) or (5.3).

In summary, using the six regular FDTD equations and 12 equations like (5.11), the NPML algorithm consists of computing the E components and ξ_e variables at times n, $n+1$, and the H components and ξ_h variables at times $n-1/2$, $n+1/2$. Notice that the ξ variables can be advanced at all the points of the grid just after the advance of the E or H corresponding components, so that no previous values of E or H need to be stored. For example ξ_{hzy} can be advanced with (5.11) just after advancing H_z, in the same space loop.

5.1.4 FDTD Scheme for the Uniaxial PML

The time domain equations of the uniaxial PML are composed of a set identical to the Maxwell equations in a lossy medium, with in addition a set of six differential equations. The D_x equation (4.19) can be discretized as follows:

$$D_x^{n+1} = A_y D_x^n + \Delta t B_y \left[\frac{\Delta H_z^{n+1/2}}{\Delta y} - \frac{\Delta H_y^{n+1/2}}{\Delta z}\right] \tag{5.12}$$

where A_y and B_y are given by (5.2) or (5.3). The differential equation (4.18) is discretized in [31] as

$$\frac{D_x^{n+1} - D_x^n}{\Delta t} + \frac{\sigma_x}{\varepsilon_0} \frac{D_x^{n+1} + D_x^n}{2} = \varepsilon_0 \frac{E_x^{n+1} - E_x^n}{\Delta t} + \sigma_z \frac{E_x^{n+1} + E_x^n}{2} \qquad (5.13)$$

that leads to

$$E_x^{n+1} = C E_x^n + \frac{1}{\varepsilon_0} \left[F D_x^{n+1} - G D_x^n \right] \qquad (5.14)$$

where

$$C = \frac{2\varepsilon_0 - \sigma_z \Delta t}{2\varepsilon_0 + \sigma_z \Delta t}; \quad F = \frac{2\varepsilon_0 + \sigma_x \Delta t}{2\varepsilon_0 + \sigma_z \Delta t}; \quad G = \frac{2\varepsilon_0 - \sigma_x \Delta t}{2\varepsilon_0 + \sigma_z \Delta t}. \qquad (5.15)$$

Assuming that E_x and auxiliary variable D_x are known at time n, and that $\boldsymbol{H}$ components are known at time $n + 1/2$, Eq. (5.12) allows D_x to be advanced to $n + 1$. Then, (5.14) provides us with E_x at time $n + 1$. Equations like (5.12) and (5.14) hold for E_y and E_z, with corresponding D_y and D_z. Finally, dual discretized equations can be derived in the same way for the advance of the $\boldsymbol{H}$ field from $n - 1/2$ to $n + 1/2$, with also three auxiliary variables and three auxiliary equations.

We notice that in the most general PML where three conductivities are different there are 12 variables to be stored and advanced in time, as with the split PML. Conversely, in the walls of a PML, where two stretching factors equal unity, only eight quantities have to be stored and advanced. Consider for instance a wall perpendicular to x, then in tensor (3.49) only s_x differs from unity so that only s_x is present in the E_y and E_z counterparts of (4.15). These two equations are then like equations in a regular lossy medium so that no auxiliary equation is needed for the advance of E_y and E_z. Finally, in the walls of a PML, an auxiliary variable is only required for the longitudinal components, that is E_x and H_x in a wall normal to x direction. In terms of memory, the requirements of the uniaxial PML are then identical to those of the split PML with the eight-variable discretization in Section 4.1.1.

5.1.5   A Comparison of the Requirements of the Different Versions of the PML ABC

The computational requirements of the different PMLs are compared in Table 5.1. The requirements are those of the wall PML that corresponds to most of the PML ABC with a 3D computational domain. Storage requirements and numbers of additions and multiplications per time step are given for one FDTD cell.

In terms of storage requirements, the uniaxial PML and the eight-variable split PML are the least demanding. In terms of computational time, multiplications are more costly than

TABLE 5.1: Number of Variables to be Stored and Number of Operations per FDTD Cell in the Walls of the Different PML Versions

	STORAGE	OPERATIONS	MULTIPLICATIONS	ADDITIONS
Regular split PML	10	56	16	40
Eight-variable split PML	8	60	18	42
Uniaxial PML	8	48	20	28
CPML	10	56	20	36
NPML	10	52	20	32

additions, so that the best two PMLs are probably the uniaxial PML and the regular split PML. Choosing the best PML in view of building a computational code may depend on such parameters as the computer where the code will be used, the encoding method, and the envisaged extensions to general media. Nevertheless, nowadays the NPML appears as a very attractive choice in many cases since it combines a moderate number of operations per cell with an easy and trivial extension to any dispersive or anisotropic medium.

5.2 FDTD SCHEMES FOR PMLS MATCHED TO LOSSY ISOTROPIC MEDIA

As in the above for the PML matched to a vacuum or any isotropic lossless medium, in the case of a PML matched to a lossy medium, various FDTD schemes can be used. They are based on the different time domain equations presented in Section 4.2.

Consider first the split PML. Equation (4.24) can be discretized easily as follows at E_x mesh nodes:

$$E_{xy}^{n+1} = A(\sigma + \sigma_y)E_{xy}^n + \frac{\Delta t}{\varepsilon_0} B(\sigma + \sigma_y) \left[\frac{\Delta(H_{zx}^{n+1/2} + H_{zy}^{n+1/2})}{\Delta y} + \frac{\sigma \sigma_y}{\varepsilon_0} S_{xy}^{n+1/2} \right] \quad (5.16)$$

where S_{xy} is an auxiliary variable computed at E_{xy} nodes, and A and B are given by (5.2) or (5.3) with $\sigma + \sigma_y$ in place of σ_u. The auxiliary variable S_{xy} is the integral in (4.24). It is advanced in time with

$$S_{xy}^{n+1/2} = S_{xy}^{n-1/2} + \Delta t E_{xy}^n. \quad (5.17)$$

Five FDTD equations like (5.16) are obtained for the advance of the other five electric subcomponents. The advance of the magnetic subcomponents is left unchanged in comparison with a PML matched to a vacuum. In the most general case, six auxiliary variables S have to be stored and advanced in time. In the walls of a PML ABC, only two S variables are present.

In the case of the CPML, Eq. (4.26) is like in a CPML matched to a vacuum (4.10) with an additional term σE_x on the left-hand side. It can be discretized as the equation in a regular lossy medium:

$$E_x^{n+1} = A(\sigma)E_x^n + \frac{\Delta t}{\varepsilon_0} B(\sigma) \left[\frac{1}{\kappa_y} \frac{\Delta H_z^{n+1/2}}{\Delta y} - \frac{1}{\kappa_z} \frac{\Delta H_y^{n+1/2}}{\Delta z} + \psi_{hzy}^{n+1/2} - \psi_{hyz}^{n+1/2} \right] \quad (5.18)$$

where A and B are (5.2) or (5.3) with σ in place of σ_u and ψ quantities are advanced in time with (5.9). Two similar equations hold for E_y and E_z. Thus, the advance of the E field in the PML is like in the corresponding nonstretched lossy medium with two ψ_h variables added to the discretized components of the H curl. And the advance of the H field is like in the CPML matched to a vacuum.

Consider now the NPML. Equation (4.27) is like in the corresponding nonstretched lossy medium, so that the FDTD advance of E_x is performed with the FDTD equation of the lossy medium, with ξ_h quantities instead of the H components in the curl. This also holds for E_y and E_z. In addition, the same equations as in a PML matched to a vacuum (5.11) are used for advancing the ξ variables.

Finally, consider the uniaxial PML. The advance of E_x component involves three equations, namely (4.32), (4.33a), (4.33b). The first one and the last one are like (4.19) and (4.18) in a PML matched to a vacuum. So, firstly D'_x is advanced using (5.12) with D'_x in place of D_x and σ in place of σ_y. Secondly D_x is advanced with the discretized form of (4.33a):

$$D_x^{n+1} = \frac{2\varepsilon_0 - \sigma_y \Delta t}{2\varepsilon_0 + \sigma_y \Delta t} D_x^n + \frac{2\varepsilon_0}{2\varepsilon_0 + \sigma_y \Delta t} \left[D'^{n+1}_x - D'^n_x \right], \quad (5.19)$$

and finally, E_x is advanced with (5.14). E_y and E_z are advanced in the same way.

5.3 FDTD SCHEMES FOR PMLS MATCHED TO ANISOTROPIC MEDIA

The split PML time domain equations (4.36) are like the split equations (3.1a), (3.1b) in a PML matched to a vacuum, with D in place of E. The FDTD equations are also like those in a vacuum, for instance of the form (5.1) for (4.36a). This allows the six subcomponents of auxiliary field D to be advanced in time, that is D^n advanced to D^{n+1}. Then, E^{n+1} is computed by solving for the E components the algebraic system (4.37). If the permittivity is also anisotropic, the advance of H field by one time step is similar, with B and H in place of D and E.

With the CPML or the NPML, obtaining the discretized equations of the corresponding PML is straightforward, provided that the time domain equations and their discretized counterparts do exist in the considered anisotropic medium. Like with a PML matched to a vacuum or to an isotropic lossy medium, described in detail in the above:

- in the CPML, the FDTD equations are the same as in the corresponding medium, with only the addition of ψ variables to the discretized curls. The ψ variables are advanced in time with (5.9), (5.10) that do not depend on the constitutive parameters of the medium.

- in the NPML, the FDTD-PML equations are the same as in the corresponding medium, with ξ variables instead of E and H components in the curls. The ξ variables are advanced with equations like (5.11) that do not depend on the constitutive parameters of the medium.

In terms of effort to encoding the scheme, the CPML and the NPML are very attractive, because most of the code used in the inner domain can be reused in the PML ABC. The only modification in the PML is the addition of the advance of the auxiliary variables ψ or ξ, and the introduction of these variables in the right-hand side of the FDTD equations.

5.4   FDTD SCHEMES FOR PMLS MATCHED TO DISPERSIVE MEDIA

In the cases of the CPML and the NPML, the comments in the previous paragraph also hold. The FDTD equations in the PML are identical to those in the dispersive medium, with only the addition of ψ variables to the curls in the CPML, and with ξ variables instead of E and H components in the curls of the NPML. So, once again, on condition that a FDTD scheme is available for the dispersive medium, deriving the FDTD scheme for the corresponding CPML or NPML is trivial, even in the case where the medium is dispersive and anisotropic.

In the case of the uniaxial PML matched to isotropic dispersive media reported in [33], using (4.45) and (4.46) three time domain equations are obtained for each component of the E field, as using (4.30) and (4.31) in the PML for lossy media. Since (4.45b) and (4.46) are like (4.30b) and (4.31), the FDTD advance of component E_x is also performed in three steps. Two steps are like in the lossy case. The third step that replaces (5.19) is governed by the time domain counterpart of (4.45a) that involves the frequency-dependent permittivity of the medium. For the discretization of this equation, the reader is referred to [33, 40].

5.5   PROFILES OF CONDUCTIVITY IN THE PML ABC

As mentioned in Section 2.4, in numerical methods a spurious reflection is produced from vacuum–PML and PML–PML interfaces. To reduce this reflection the conductivity in the PML must grow from a small value in the vacuum–PML interface to a larger value beside the PEC that ends the PML. Two profiles of conductivity are of current use, namely the polynomial profile and the geometrical profile [41]. Only the electric conductivity is considered in the following. The magnetic conductivity can be obtained using the matching condition (1.9).

With the polynomial profile, the PML conductivity σ_ρ, either σ_x, σ_y, or σ_z, varies as

$$\sigma_\rho(\rho) = \sigma_m \left(\frac{\rho}{\delta}\right)^n \qquad (5.20)$$

where n is the degree of the polynomial, δ is the PML thickness, ρ is the distance from the interface, and σ_m is the conductivity on the outer side of the PML (for $\rho = \delta$). Using (2.26b) the theoretical reflection at normal incidence is then

$$R(0) = e^{-(2/(n+1))(\sigma_m \delta / \varepsilon_0 c)}. \qquad (5.21)$$

From this, for a given reflection $R(0)$, σ_m reads

$$\sigma_m = -\frac{(n+1)\varepsilon_0 c}{2\delta} \ln R(0). \qquad (5.22)$$

The actual conductivity implemented at FDTD mesh node of index L ($L = 0$ in the interface) is computed as

$$\sigma_\rho(L) = \frac{1}{\Delta\rho} \int_{\rho(L)-\Delta\rho/2}^{\rho(L)+\Delta\rho/2} \sigma_\rho(u)du \qquad (5.23)$$

where $\Delta\rho$ is the space step. At mesh points of indexes $L = 0, 1/2, 1, \ldots, N-1/2$, this leads to

$$\sigma_\rho(0) = \frac{\sigma_m}{(n+1)2^{n+1} N^n} = -\frac{\varepsilon_0 c \ln R(0)}{2^{n+2} \Delta\rho N^{n+1}} \qquad (5.24a)$$

$$\sigma_\rho(L > 0) = \sigma_\rho(0) \left[(2L+1)^{n+1} - (2L-1)^{n+1}\right] \qquad (5.24b)$$

where N is the number of FDTD cells in the PML thickness.

With a geometrical profile the conductivity grows as a geometrical progression of the form

$$\sigma_\rho(\rho) = \sigma_0 \left(g^{1/\Delta\rho}\right)^\rho \qquad (5.25)$$

so that the conductivity is multiplied with factor g from one FDTD cell to the next. The corresponding normal reflection (2.26b) is then

$$R(0) = e^{-(2/\varepsilon_0 c)[(g^N - 1)/\ln g]\sigma_0 \Delta\rho}, \qquad (5.26)$$

and for a given reflection $R(0)$, σ_0 reads

$$\sigma_0 = -\frac{\varepsilon_0 c}{2\Delta\rho} \frac{\ln g}{g^N - 1} \ln R(0). \tag{5.27}$$

Finally, using (5.23) the numerical conductivity at indexes $L = 0, 1/2, 1, \ldots, N - 1/2$, reads

$$\sigma_\rho(0) = \sigma_0 \frac{\sqrt{g} - 1}{\ln g} = \frac{\varepsilon_0 c(1 - \sqrt{g}) \ln R(0)}{2\Delta\rho(g^N - 1)} \tag{5.28a}$$

$$\sigma_\rho(L > 0) = \sigma_0 \frac{g - 1}{\sqrt{g} \ln g} g^L = \frac{\varepsilon_0 c(1 - g) \ln R(0)}{2\Delta\rho \sqrt{g}(g^N - 1)} g^L. \tag{5.28b}$$

The polynomial profile is the most used in the literature, with degrees ranging from 2 to 5. The geometrical profile has an interesting property. The ratio of successive numerical conductivities (5.28b) is constant throughout the PML. Since a PML with a growing conductivity is nothing but a juxtaposition of successive PMLs with different conductivities, a spurious reflection occurs from every inner interface. This reflection mainly depends on the ratio of the successive conductivities. From this, the constant ratio of the geometrical profile is better suited to reduce the spurious reflection. This is especially the case when evanescent waves are present. In that case, the geometrical profile allows the PML thickness to be thinner than with the polynomial profile. This was investigated in [15, 41], and is discussed in Chapter 6.

5.6   THE PML ABC IN THE DISCRETIZED FDTD SPACE

In the previous chapters, the properties of PML media are derived in the continuous space. In view of application as an ABC, the most important feature is the absence of reflection from the interface between an inner medium and its corresponding PML. Unfortunately, in the discretized space of the FDTD method, or more generally with any finite method, things are a little different. An amount of reflection is produced from vacuum–PML, or from PML–PML, interfaces, despite the fact that the continuous theory predicts no reflection. It can be shown easily by means of numerical experiments that this spurious reflection may be either quite small, smaller than -80 dB for example, or quite large, close to total reflection. It also appears that the reflection strongly depends on parameters such as the thickness of the PML expressed in cells, the conductivity profile $\sigma_\rho(\rho)$ in the PML, or the separation between the PML and the object of interest. All this implies that a theory is needed in view of predicting the spurious reflection, so as to be able to control it and to optimize the parameters of the PML ABC.

A number of experiments showing the presence of numerical reflection can be found in the literature. Here we show three experiments drawn out from [5], [41], and [42]. The first one deals with the reflection of traveling plane waves from a plane PML boundary like the one in Fig. 2.4. Details of the calculation can be found in [5]. Results in Fig. 5.1 give the theoretical

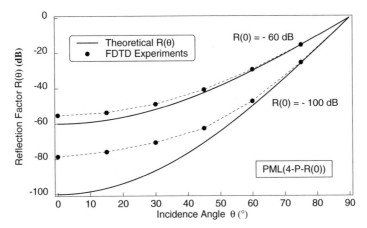

FIGURE 5.1: Comparison of the theoretical reflection $R(\theta)$ with results of FDTD experiments from [5], for $R(0) = -60$ dB and $R(0) = -100$ dB. The PML is four cells in thickness with a parabolic profile of conductivity ($n = 2$ in (5.20))

reflection (2.26), and the observed reflection from FDTD simulations, for two values of the normal reflection (2.26b). As observed, the actual FDTD reflection significantly differs from the theoretical reflection. Notice that the smaller the theoretical reflection $R(0)$, the larger the difference between the actual value and the theoretical value.

The second experiment is provided in Fig. 5.2. In that case, a 20-cell-long thin plate is placed within a domain surrounded with a PML ABC. A unit-step incident wave strikes the plate. What is reported in Fig. 5.2 is the field normal to the surface of the plate at its end. Besides a reference solution computed within a very large domain, denoted as the exact solution, results are given for various plate–PML separations, from 2 FDTD cells to 20 FDTD cells. As clearly observed, the results strongly depart from the reference solution when the PML–plate separation is small. From the discussion in Chapter 1 about the location of ABCs with respect to the sources, such a behavior in function of the plate–PML separation suggests that the PML does not absorb properly the evanescent fields that surround the plate. This will be confirmed later in this chapter by means of the theory of the numerical reflection.

The last experiment shows the field radiated from a short dipole antenna. The upper part of Fig. 5.3 compares the fields computed using Higdon operators, the matched layer ABC [13, 14], and three PML ABCs of thicknesses 4, 8, 12 FDTD cells, respectively. As observed, the PML ABC yields about perfect results, even with a 4-cell-thick PML. Nevertheless, in frequency domain it appears that the results strongly depart from the analytical solution below a certain frequency with the 4- and 8-cell-thick PMLs. This will be interpreted later with the theory of the numerical reflection.

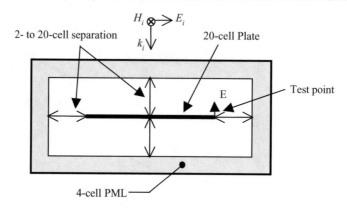

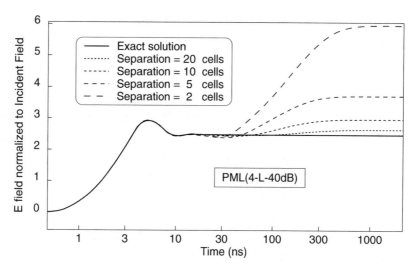

FIGURE 5.2: Electric field normal to the surface of a 20-cell-long 2D thin plate. The space step is 5 cm, the PML conductivity is linear ($n = 1$ in (5.20)), and the incident wave is of the form $[1 - \exp(t/\tau)]$, where $\tau = 1$ ns

In the following, the theory of the propagation and reflection of waves in the FDTD PML is derived in the 2D case. As mentioned in Chapter 2, use of a PML medium as an ABC is mainly a 2D problem, even in 3D applications. Moreover, in realistic applications of the FDTD method, the theory cannot predict with great accuracy the reflection from a PML ABC, because the waves that will strike the ABC are not known before the calculation. The objective of the numerical theory is only to allow us to understand why an amount of spurious reflection is present, so as to be able to design the parameters of the PML in view of reducing its thickness and placing it as close as possible to the region of interest of the computational domain. To this end, the 2D case is sufficient.

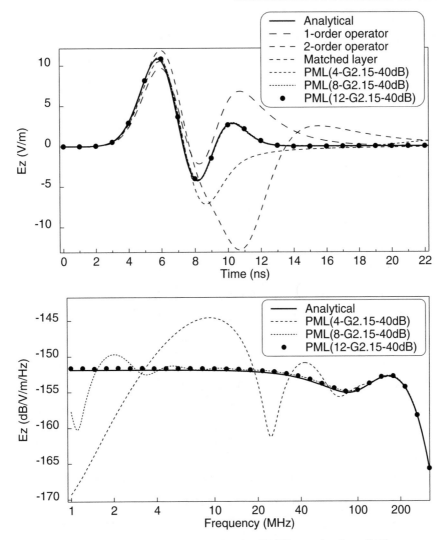

FIGURE 5.3: Electric field radiated by a short dipole. FDTD results from [42] are compared with analytical results in time domain (upper part) and in frequency domain (lower part). The conductivity in the PMLs varies geometrically (5.25) with $g = 2.15$ and $R(0) = -40$ dB. The dipole and the point of observation are 2 FDTD cells from two corners of a 14-14-14-cell vacuum surrounded with the ABCs

5.6.1 Propagation of Plane Waves in the Split FDTD-PML

In this paragraph, we consider the split PML, in the 2D case of Chapter 2. The CPML, the NPML, and the uniaxial PML will be addressed later. The stretching factor is (2.6), the case of the CFS-PML factor (3.50) will also be addressed later.

Let us consider the 2D FDTD equations corresponding to the continuous equations (2.3), in the case where the matching condition (1.9) holds:

$$E_x^{n+1} = A_y\, E_x^n + \frac{\Delta t}{\varepsilon_0}\, B_y\, \frac{\Delta H_{zx}^{n+1/2} + \Delta H_{zy}^{n+1/2}}{\Delta y} \qquad (5.29a)$$

$$E_y^{n+1} = A_x\, E_y^n - \frac{\Delta t}{\varepsilon_0}\, B_x\, \frac{\Delta H_{zx}^{n+1/2} + \Delta H_{zy}^{n+1/2}}{\Delta x} \qquad (5.29b)$$

$$H_{zx}^{n+1/2} = A_x H_{zx}^{n-1/2} - \frac{\Delta t}{\mu_0}\, B_x\, \frac{\Delta E_y^n}{\Delta x} \qquad (5.29c)$$

$$H_{zy}^{n+1/2} = A_y\, H_{zy}^{n-1/2} + \frac{\Delta t}{\mu_0}\, B_y\, \frac{\Delta E_x^n}{\Delta y} \qquad (5.29d)$$

with A and B from (5.2) or (5.3). By enforcing a plane wave of the form (2.4) in (5.29), we obtain

$$\varepsilon_0 \Sigma_y E_{0x} = -\frac{\Delta t}{\Delta y}\, B_y \sin \frac{\overline{k}_y \Delta y}{2}\, H_{0z} \qquad (5.30a)$$

$$\varepsilon_0 \Sigma_x E_{0y} = \frac{\Delta t}{\Delta x}\, B_x \sin \frac{\overline{k}_x \Delta x}{2}\, H_{0z} \qquad (5.30b)$$

$$\mu_0 \Sigma_x H_{0zx} = \frac{\Delta t}{\Delta x}\, B_x \sin \frac{\overline{k}_x \Delta x}{2}\, E_{0y} \qquad (5.30c)$$

$$\mu_0 \Sigma_y H_{0zy} = -\frac{\Delta t}{\Delta y}\, B_y \sin \frac{\overline{k}_y \Delta y}{2}\, E_{0x} \qquad (5.30d)$$

where $\overline{k}_x$ and $\overline{k}_y$ are the components of the wave vector in the discretized space, and:

$$\Sigma_u = \frac{e^{j\omega \Delta t/2} - A_u e^{-j\omega \Delta t/2}}{2j} \qquad (u = x, y). \qquad (5.31)$$

System (5.30) is identical to (2.5) with the changes

$$k_u \rightarrow \frac{1}{\Delta u} \sin \frac{\overline{k}_u \Delta u}{2} \qquad (u = x, y) \qquad (5.32a)$$

$$s_u = s_u^* \rightarrow \frac{\Sigma_u}{\omega \Delta t} \frac{1}{B_u} \qquad (u = x, y) \qquad (5.32b)$$

so that the solutions of (5.30) can be obtained from the solutions of (2.5) with the changes (5.32). This yields the following equation of dispersion in the FDTD PML,

$$\frac{1}{c^2 \Delta t^2} = \frac{1}{\Omega_x^2 \Delta x^2} \sin^2 \frac{k_x \Delta x}{2} + \frac{1}{\Omega_y^2 \Delta y^2} \sin^2 \frac{k_y \Delta y}{2} \qquad (5.33)$$

where

$$\Omega_u = \Sigma_u / B_u \quad (u = x, y), \tag{5.34}$$

and the following wave numbers, in the general case of nonhomogeneous waves:

$$\sin \frac{\overline{k}_x \Delta x}{2} = \frac{\Delta x}{c \Delta t} \Omega_x C(\chi, \theta) \tag{5.35a}$$

$$\sin \frac{\overline{k}_y \Delta y}{2} = \frac{\Delta y}{c \Delta t} \Omega_y S(\chi, \theta). \tag{5.35b}$$

Notice that (5.35) yields (2.27) if the space and time steps vanish. Using (5.32) in (2.12) the ratio E_0/H_0 (2.14) is left unchanged, it equals the impedance of a vacuum, as in the continuous PML.

5.6.2 Reflection from a PML–PML Interface

Let us now consider the transmission and reflection of a wave at an interface between two infinite PMLs, with uniform conductivities σ_{x1} and σ_{x2} in the PMLs and σ_{x0} in the interface (Fig. 5.4). As in the continuous case, the $\overline{k}_y$ wave numbers in the two media are equal, that is $\overline{k}_{y1} = \overline{k}_{yr} = \overline{k}_{y2}$. From this and from (5.35b), the set (2.33) also holds. Because in the FDTD grid there are E_y nodes in the interface, the same E_y values are used in the two media for the advance of field components half a cell from the interface, so that continuity of E_y is retained in the interface. This reads $E_{y1} + E_{yr} = E_{y2}$, or with $x = 0$ in the interface, $E_{0y1} + E_{0yr} = E_{0y2}$,

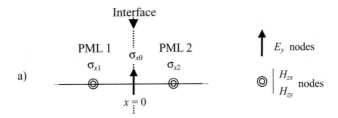

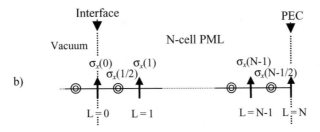

FIGURE 5.4: An interface between, (a) two infinite PML media of conductivities σ_{x1} and σ_{x2}, (b) a vacuum and a N-cell-thick PML ABC with conductivity growing in the PML

as in the continuous case (2.18a). There is no H_z node in the interface. In place of (2.18b), a second equation connecting the incident, reflected, and transmitted waves is provided with the FDTD equation of the interface (5.29b)

$$E_y^{n+1}(x=0) = A_{x0} E_y^n(x=0)$$

$$+ \frac{\Delta t}{\varepsilon_0} \frac{B_{x0}}{\Delta x} \left[H_z^{n+1/2} \left(x = \frac{\Delta x}{2} \right) - H_z^{n+1/2} \left(x = -\frac{\Delta x}{2} \right) \right] \quad (5.36)$$

where A_{x0} and B_{x0} are (5.2), (5.3) for conductivity σ_{x0}. Consider now components E_y and H_z. At location x in medium 1, with $x = 0$ in the interface they can be written as

$$E_{y1} + E_{yr} = \left[E_{0y1} e^{-j\bar{k}_{x1}x} + E_{0yr} e^{-j\bar{k}_{xr}x} \right] e^{j\omega t - j\bar{k}_{y1}y} \quad (5.37a)$$

$$H_{z1} + H_{zr} = \left[H_{0z1} e^{-j\bar{k}_{x1}x} + H_{0zr} e^{-j\bar{k}_{xr}x} \right] e^{j\omega t - j\bar{k}_{y1}y}, \quad (5.37b)$$

and in medium 2:

$$E_{y2} = E_{0y2} e^{j\omega t - j\bar{k}_{x2}x - j\bar{k}_{y1}y} \quad (5.38a)$$

$$H_{z2} = H_{0z2} e^{j\omega t - j\bar{k}_{x2}x - j\bar{k}_{y1}y}. \quad (5.38b)$$

Moreover, using (5.30b), (5.34), and (5.35a), H_{0z} can be expressed as

$$H_{0z} = \sqrt{\frac{\varepsilon_0}{\mu_0}} \frac{1}{C(\chi, \theta)} E_{0y}, \quad (5.39)$$

so that H_{0z1}, H_{0zr}, H_{0z2} can be replaced with E_{0y1}, E_{0yr}, E_{0y2}, in (5.37b) and (5.38b). Defining then R as the ratio of the reflected to the incident E fields in the interface, that is $R = E_{0yr}/E_{0y1}$, we have $E_{0yr} = R E_{0y1}$ and from the continuity of E_y in the interface, $E_{0y2} = (1 + R) E_{0y1}$. Finally, since with (2.33) we have $\bar{k}_{xr} = -\bar{k}_{x1}$ and $C(\chi_1, \theta_1) = -C(\chi_r, \theta_r) = C(\chi_2, \theta_2)$, Eqs. (5.37) and (5.38) become

$$E_{y1} + E_{yr} = E_{0y1} \left[e^{-j\bar{k}_{x1}x} + R e^{j\bar{k}_{x1}x} \right] e^{j\omega t - j\bar{k}_{y1}y} \quad (5.40a)$$

$$H_{z1} + H_{zr} = \sqrt{\frac{\varepsilon_0}{\mu_0}} \frac{1}{C(\chi_1, \theta_1)} E_{0y1} \left[e^{-j\bar{k}_{x1}x} - R e^{j\bar{k}_{x1}x} \right] e^{j\omega t - j\bar{k}_{y1}y} \quad (5.40b)$$

$$E_{y2} = (1 + R) E_{0y1} e^{j\omega t - j\bar{k}_{x2}x - j\bar{k}_{y1}y} \quad (5.41a)$$

$$H_{z2} = (1 + R) \sqrt{\frac{\varepsilon_0}{\mu_0}} \frac{1}{C(\chi_1, \theta_1)} E_{0y1} e^{j\omega t - j\bar{k}_{x2}x - j\bar{k}_{y1}y}. \quad (5.41b)$$

Using these field components in (5.36), we obtain

$$(1 + R)e^{j\omega\Delta t/2} = A_{x0}(1 + R)e^{-j\omega\Delta t/2}$$

$$- \frac{\Delta t}{\sqrt{\varepsilon_0\mu_0}\Delta x} B_{x0} \frac{1}{C(\chi_1, \theta_1)} \left[(1 + R)e^{-j\bar{k}_{x2}\Delta x/2} - (e^{j\bar{k}_{x1}\Delta x/2} - Re^{-j\bar{k}_{x1}\Delta x/2}) \right]. \quad (5.42)$$

By solving for R, the reflection from the PML–PML interface is then

$$R = -\frac{2j\Omega_{x0} + \frac{c\Delta t}{\Delta x}\frac{1}{C(\chi_1,\theta_1)} \left[e^{-j\bar{k}_{x2}\Delta x/2} - e^{j\bar{k}_{x1}\Delta x/2} \right]}{2j\Omega_{x0} + \frac{c\Delta t}{\Delta x}\frac{1}{C(\chi_1,\theta_1)} \left[e^{-j\bar{k}_{x2}\Delta x/2} + e^{-j\bar{k}_{x1}\Delta x/2} \right]}. \quad (5.43)$$

The exponentials involving k_x wave numbers can be removed using (5.35a). This leads to

$$R = -\frac{2j\Omega_{x0} + \Lambda_{x2} - j\Omega_{x2} - \Lambda_{x1} - j\Omega_{x1}}{2j\Omega_{x0} + \Lambda_{x2} - j\Omega_{x2} + \Lambda_{x1} - j\Omega_{x1}} \quad (5.44)$$

where

$$\Lambda_{xi} = \sqrt{\frac{1}{C(\chi_1, \theta_1)^2}\frac{c^2\Delta t^2}{\Delta x^2} - \Omega_{xi}^2}. \quad (5.45)$$

Notice that R depends on conductivities σ_{x1}, σ_{x2}, σ_{x0} from Ω_{x1}, Ω_{x2}, Ω_{x0} (5.34). It also depends on the incidence angle θ_1 and on the evanescence parameter χ_1 from Λ_{x1}, Λ_{x2} (5.45). As can be shown, (5.43) holds as a special case ($\chi_1 = 1$, so that $C(\chi_1, \theta_1) = \cos\theta_1$) the reflection coefficient given in [43] and [44]. A software package for computing reflection (5.44) is provided at www.morganclaypool.com/page/berenger.

An important simplification of (5.44) is obtained in the case where the parameter $\cosh\chi_1$ of the incident wave is large enough (strongly evanescent waves). In this case $\Lambda_{xi} \to j\Omega_{xi}$ so that R tends to

$$R_\infty = -\frac{\Omega_{x0} - \Omega_{x1}}{\Omega_{x0}}. \quad (5.46)$$

Moreover, by assuming that $\omega\Delta t \ll 1$ and $\sigma_x\Delta t/\varepsilon_0 \ll 1$ for all the conductivities, two assumptions that usually hold in PMLs, the Ω_{xi} parameters can be approximated as

$$\Omega_{xi} \approx \frac{1}{2j}(j\omega\Delta t + \sigma_{xi}\Delta t/\varepsilon_0) \quad (5.47)$$

so that (5.46) becomes

$$R_\infty = \frac{\sigma_{x1} - \sigma_{x0}}{j\omega\varepsilon_0 + \sigma_{x0}}. \quad (5.48)$$

Finally, at a vacuum–PML interface ($\sigma_{x1} = 0$) the limit of R as cosh χ_1 tends to infinity reads

$$R_\infty = \frac{j\sigma_{x0}/\varepsilon_0\,\omega}{1 - j\sigma_{x0}/\varepsilon_0\,\omega}. \qquad (5.49)$$

An example of reflection R from (5.44) is shown in Fig. 5.5, for the incidence $\theta_1 = \pm\pi/4$ and cosh χ_1 in the range 1–1000 ($\theta_1 = +\pi/4$ and $\chi_1 < 0$, or $\theta_1 = -\pi/4$ and $\chi_1 > 0$, so that the wave propagates and is evanescent toward $+x$). As observed, from traveling waves (cosh $\chi_1 = 1$) to strongly evanescent waves (large cosh χ_1) R grows up to R_∞. In accordance with (5.49), the strongly evanescent waves are reflected in totality ($R = -1$) at frequencies far smaller than cutoff

$$f_c = \frac{\sigma_{x0}}{2\pi\,\varepsilon_0} \qquad (5.50)$$

that equals 10 MHz in the case of Fig. 5.5.

Validity of (5.46)–(5.48) is that of the assumption $\Lambda_{xi} \to j\,\Omega_{xi}$, for $i = 1, 2$. Using (5.47) and with $\chi_1 = \chi_2$ and $\theta_1 = \theta_2$ from (2.23), this yields the condition

$$|C(\chi_1, \theta_1)| \gg \frac{c\,\Delta t}{\Delta x}\left|\frac{2j}{j\omega\Delta t + \sigma_{xi}\Delta t/\varepsilon_0}\right| \quad (i = 1, 2). \qquad (5.51)$$

At low frequency, for $f \ll \sigma_{xi}/2\pi\,\varepsilon_0$, (5.51) reduces to

$$\cosh \chi_1 \gg \frac{2c\,\varepsilon_0}{\sigma_{xi}\Delta x} \quad (i = 1, 2). \qquad (5.52)$$

Condition (5.52) can be interpreted by considering the absorption coefficient of a plane wave in a continuous PML (2.35). If (5.52) holds the exponential coefficient in (2.35) is close to zero for $x = \Delta x$. This means that the evanescent wave in the PMLs must be absorbed in totality upon a range shorter than the FDTD cell size Δx, i.e., within one cell. This is summarized in Fig. 5.6. This cannot be achieved, resulting in a strong numerical reflection from the interface. Notice that in the case of an interface between a vacuum ($\sigma_{x1} = 0$) and a PML, (5.52) requires cosh χ_1 to be infinite. In fact, in that special case, using (5.47) in (5.44) it can be shown easily that a sufficient condition for the reflection to be total below frequency f_c (5.50) is that (5.52) holds for conductivity σ_{x0} (and for σ_{x2} because $\sigma_{x0} < \sigma_{x2}$). In the general case, for $f > \sigma_{xi}/2\pi\,\varepsilon_0$ (5.51) shows that smaller values of cosh χ_1 make R_∞ valid. This also holds with a vacuum–PML interface for $f > f_c$, as observed in Fig. 5.5.

Let us now consider the case of homogeneous waves, where $C(\chi_1, \theta_1) = \cos\theta_1$. Assuming as in the above that $\omega_0\Delta t \ll 1$ and $\sigma_x \Delta t/\varepsilon_0 \ll 1$, the following approximation to (5.44) can

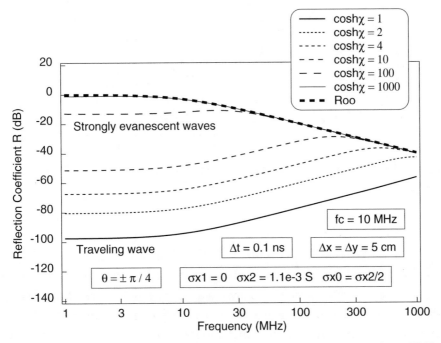

FIGURE 5.5: Reflection (5.44) from the interface between a vacuum and an infinite PML medium, for incidence $\theta = \pm\pi/4$ and various values of the evanescence coefficient $\cosh\chi$, from traveling waves ($\cosh\chi = 1$) to strongly evanescent waves (large $\cosh\chi$)

be derived:

$$R \approx -\frac{\cos\theta_1}{4\varepsilon_0 c}(2\sigma_{x0} - \sigma_{x1} - \sigma_{x2})\Delta x - \frac{\cos^2\theta_1}{16\varepsilon_0^2 c^2}(\sigma_{x1} + \sigma_{x2} + 2j\omega\varepsilon_0)(\sigma_{x2} - \sigma_{x1})\Delta x^2. \quad (5.53)$$

The first-order term does not depend on frequency. At an interface between a vacuum and a PML of constant conductivity ($\sigma_{x1} = 0$ and $\sigma_{x0} = \sigma_{x2}$), this term is proportional to the PML conductivity σ_{x2}. In the case where $\sigma_{x0} = (\sigma_{x1} + \sigma_{x2})/2$ the first-order term vanishes and the reflection is proportional to Δx^2, as noted in [43]. With σ_{x1}, σ_{x2}, σ_{x0}, Δx, θ in Fig. 5.5, (5.53) yields $R = -97.46$ dB for $\omega\varepsilon_0 \ll \sigma_{x2}$, in accordance with the low frequency plateau of the traveling wave curve in the figure.

In summary, at a single interface between two infinite PMLs the refection of homogeneous traveling waves mainly depends on the difference of the two conductivities. At a vacuum–PML interface the reflection can be reduced by decreasing the conductivity in the PML. But for a PML of finite thickness the overall absorption is then decreased. This is why a conductivity growing from a small value in the vacuum–PML interface to a larger value beside

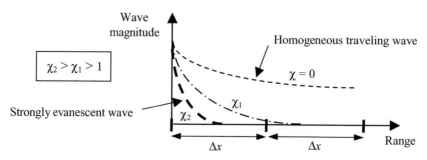

FIGURE 5.6: Theoretical decrease of a wave in the PML medium versus the FDTD grid, for various evanescence parameters χ

the outer side of the PML is used in actual FDTD PMLs. The reflection coefficient for such PMLs of finite thickness is derived in the next paragraph. The most important conclusion of the present paragraph is that evanescent waves may be strongly reflected from FDTD PMLs. This reflection is total below frequency (5.50) if (5.52) holds. This is the case in Fig. 5.2 where the low frequencies are strongly reflected when the PML is close to the plate, that is within the evanescent region, resulting in a late time spurious reflection in time domain. Such a strong reflection of evanescent waves is present in most physical problems. From this, one could conclude that the PML ABC suffers from the same drawback as the previously used ABCs that could not be placed within the evanescent regions. In fact, some remedies discussed later will render the PML ABC capable of absorbing evanescent waves with a relatively small spurious reflection, so that it can be placed within evanescent regions, close to the region of interest of actual calculations.

5.6.3   Reflection from a N-Cell-Thick PML

We now consider an incident wave with parameters χ_1, θ_1, and a PML N cells in thickness with a nonuniform conductivity, i.e., a conductivity depending on the mesh index L (Fig. 5.4). The wave transmitted into the PML is of the form (2.12), (2.13), with the sine and cosine replaced with (2.28). At every interface, $\bar{k}_y$ is left unchanged and (2.33) holds, so that $\bar{k}_y = \bar{k}_{y1}$, $\chi = \chi_1$, and $\theta = \theta_1$ in the whole PML. The incident and reflected waves can be written as with a single interface (5.37), (5.40). Denoting by $T(L)$ an unknown quantity at row L, let the electric field E_y be written as

$$E_y(L) = E_{0y1} T(L) e^{j\omega t - j\bar{k}_{y1}y} \quad (L = 1, \ldots, N-1). \tag{5.54}$$

Similarly, at row $L + 1/2$, using (2.12), (2.13) and (5.39), let the magnetic field be written as

$$H_z(L + 1/2) = \sqrt{\frac{\varepsilon_0}{\mu_0}} E_{0y1} \frac{1}{C(\chi_1, \theta_1)} T(L + 1/2) e^{j\omega t - j\bar{k}_{y1} y} \quad (L = 0, \dots, N - 1) \quad (5.55a)$$

$$H_{zx}(L + 1/2) = C(\chi_1, \theta_1)^2 H_z(L + 1/2) \quad (L = 0, \dots, N - 1). \quad (5.55b)$$

And finally, at the end of the PML

$$E_y(N) = 0. \quad (5.56)$$

Using (5.40a), (5.54) and (5.55) into the N FDTD equations of the advance on time of E_y, and into the N equations of the advance of H_{zx} (5.29c), from $L = 0$ to $L = N - 1$, we obtain a set of $2N$ equations for the $2N$ unknowns R, $T(1/2), \dots, T(N - 1/2)$. After eliminating the incident wave number using (5.35a), this set can be written in the form

$$M \cdot \begin{pmatrix} R \\ T(1/2) \\ T(1) \\ \cdot \\ \cdot \\ T(N - 1/2) \end{pmatrix} = \begin{pmatrix} V \\ \alpha D(1/2) \\ 0 \\ \cdot \\ \cdot \\ 0 \end{pmatrix} \quad (5.57)$$

where M is the tridiagonal matrix

$$M = \begin{pmatrix} U & \alpha D(0)\dots\dots\dots\dots\dots\dots\dots\dots\dots \\ \dots\dots\dots\dots\dots\dots\dots\dots\dots\dots \\ \dots\dots -\alpha D(L) & 1 & \alpha D(L)\dots\dots\dots\dots\dots \\ \dots\dots -\alpha D(L + 1/2) & 1 & \alpha D(L + 1/2)\dots\dots \\ \dots\dots\dots\dots\dots\dots\dots\dots\dots \\ \dots\dots\dots\dots\dots\dots -\alpha D(N - 1/2) & 1 \end{pmatrix}$$

and:

$$\alpha = \frac{c \Delta t}{\Delta x} \frac{1}{C(\chi_1, \theta_1)}; \quad D(L) = \frac{1}{2j\Omega_x(L)}$$

$$U = 1 + \alpha D(0) \left[\sqrt{1 - Q^2} - jQ \right]; \quad Q = \frac{1}{\alpha} \sin \frac{\omega \Delta t}{2}$$

$$V = -1 + \alpha D(0) \left[\sqrt{1 - Q^2} + jQ \right].$$

System (5.57) can be solved recursively for the unknown of interest R. Notice that the reflection depends on the conductivity profile $\sigma_x(L)$ in the PML through $\Omega_x(L)$.

For strongly evanescent waves, i.e., if $\cosh \chi_1$ is large enough, α vanishes and then all the nondiagonal terms of matrix M vanish, and finally (5.57) reduces to the scalar equation $RU = V$. Under the same conditions as in the previous paragraph ($\omega \Delta t \ll 1$ and $\sigma_x(0) \, \Delta t / \varepsilon_0 \ll 1$) reflection R from (5.57) also tends to the limit R_∞ (5.49), with just $\sigma_x(0)$ in place of σ_{x0}. The reflection is total below frequency $\sigma_x(0)/2\pi \varepsilon_0$, on condition that (5.52) holds for $\sigma_x(0)$.

In conclusion, the FDTD reflection from a vacuum–PML interface can be predicted easily with (5.57) for any N-cell-thick PML with conductivity varying in the PML. To do this, a software package is provided at www.morganclaypool.com/page/berenger. What is of primordial importance is that the strongly evanescent waves are totally reflected below frequency (5.50) that is proportional to the conductivity in the interface $\sigma_x(0)$. Such a reflection is observed clearly with the field radiated from a dipole in the lower part of Fig. 5.3, where frequency f_c (5.50) equals 50.3 MHz with the 4-cell PML and 2.25 MHz with the 8-cell PML. In time domain this means that a total reflection will appear after a delay t_c of the order of the inverse of f_c (5.50). This is exactly what is observed in Fig. 5.2. Thus, the conductivity in the interface $\sigma_x(0)$ is a critical parameter as designing a PML ABC, especially in problems of interaction of an incident wave with a scattering structure that are discussed in the next chapter.

Fig. 5.7 shows a comparison of the continuous reflection (2.26a) with the discrete reflection (5.57) for the same PMLs as in Fig. 5.1. The agreement of Fig. 5.1 with Fig. 5.7, that is of FDTD experiments with reflection (5.57), is good. The small difference is due to uncertainties in the measurements of the FDTD reflection in [5].

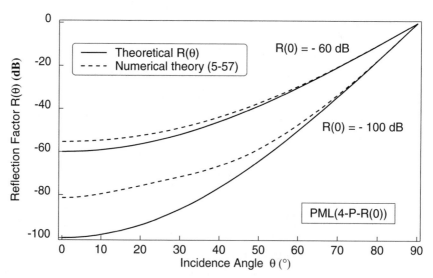

FIGURE 5.7: Comparison of the theoretical reflection $R(\theta)$ with low frequency limit of the reflection of traveling waves (5.57), for $R(0) = -60$ dB and $R(0) = -100$ dB

5.6.4   Reflection from the CPML, the NPML, and the Uniaxial PML

The numerical reflection R in the 2D TE case is derived for the uniaxial PML and the CPML in [25], and for the NPML in [29]. The derivations are close to that of the split PML in the previous paragraph, details can be found in the two papers. The results are as follows:

- the numerical reflection from the uniaxial PML and from the NPML are identical, rigorously, to that from the split PML. This means that the uniaxial PML and NPML reflections are also given by (5.44) for a single interface between two infinite PMLs, and by (5.57) for a PML of finite size like the ones used as an ABC in the FDTD method. The equality is rigorous on condition that the same discretization, either a linear one (5.2) or an exponential one (5.3), is used in the three PMLs.

- the numerical reflection from a convolutional PML (CPML) is also given by (5.44) and (5.57), but with a little change in two parameters. Coefficients A_u and B_u defined in (5.2)–(5.3) and used both in (5.44) and in (5.57) through Ω_u (5.34) must be replaced with:

$$A_u = 1; \quad B_u = \frac{1}{\kappa_u} + \frac{q_u}{1 - p_u e^{-j\omega\Delta t}} \qquad (5.58)$$

where p_u and q_u are given by (5.10) and κ_u is a real stretch. B_u in (5.58) is valid for a CFS-PML with stretching factor (3.50) and for a regular stretch (2.6) with $\kappa_u = 1$ and $\alpha_u = 0$. Then, even with a regular stretch the reflection from the CPML differs from that of the split PML, the uniaxial PML, or the NPML. From a few experiments provided in [25] it seems that the reflection of homogeneous waves from the CPML is slightly smaller than that from the other three PML implementations. Nevertheless, the difference is not significant. The interest of the CPML implementation is mainly in using a CFS stretching factor (3.50) in place of (2.6).

5.6.5   Reflection from the CFS-PML

The complex frequency shifted PML (CFS-PML) consists in using the stretching factor (3.50) in place of the regular factor (2.6). The factor (3.50) can be used with the uniaxial PML [45] or with the NPML, but the most used implementation is the convolutional CFS-PML [27]. As stated in the previous paragraph, in that case the numerical reflection is given by (5.44) and (5.57) with (5.58).

Let us now consider the case where the parameter $\cosh \chi_1$ of the incident wave is large (strongly evanescent waves). By a derivation similar to that performed with the regular PML

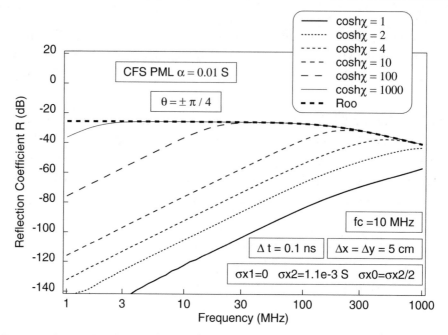

FIGURE 5.8: Reflection (5.44) from the interface between a vacuum and an infinite CFS PML medium, for various values of the evanescence coefficient cosh χ, from traveling waves (cosh $\chi = 1$) to strongly evanescent waves (large cosh χ)

in Section 5.6.2, it can be shown easily that reflection (5.44) tends to

$$R_\infty = \frac{j\frac{\alpha_{x0}+\sigma_{x0}}{\varepsilon_0\omega}}{1 - j\frac{\alpha_{x0}+\sigma_{x0}}{\varepsilon_0\omega}} \frac{\sigma_{x0}}{\alpha_{x0} + \sigma_{x0}} \qquad (5.59)$$

where σ_{x0} and α_{x0} are σ_x and α_x in the interface. An example of reflection R from (5.44) with (5.58) is shown in Fig. 5.8 for the same PMLs as in Fig 5.5. This figure clearly shows that the total reflection of strongly evanescent waves can be removed with the CFS-PML. The reflection of low frequencies is bounded with a limit R_∞ given by the second ratio in (5.59). In addition, notice that the reflection departs from R_∞ below a certain frequency. This is because the theoretical absorption decreases below f_α (3.51), resulting in a smaller numerical reflection.

Fig. 5.9 shows another comparison of the regular PML with the CFS PML [25]. The upper part shows that the CFS reflection can be widely smaller than the regular one around and below f_α (=18 MHz). However, the CFS reflection grows at low frequency, in the region where the coefficient in (3.54) grows and becomes close to unity. Thus, for the absorption of

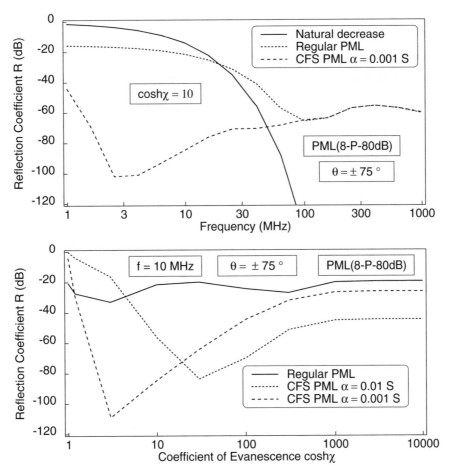

FIGURE 5.9: Reflection of evanescent waves (5.57) from PMLs. Comparison of the regular PML with the CFS PML, in function of frequency for $\cosh \chi = 10$ (upper part), and in function of $\cosh \chi$ for frequency 10 MHz (lower part)

evanescent waves, the CFS PML is only better in a band of frequency. This is in accordance with (3.54). Because the coefficient in (3.54) only depends on the product $\omega \sinh \chi$, the CFS PML can also improve the absorption in a band of $\cosh \chi$, as illustrated in the lower part of Fig. 5.9.

In summary to the results in Figs. 5.8 and 5.9, the CFS PML can widely improve the absorption of evanescent fields. Nevertheless, this improvement depends on many parameters, namely ω, $\sinh \chi$, σ_x/α_x, and θ. Especially, ω and $\sinh \chi$ are critical parameters in (3.54), so that one could fear that wide-band applications could not be achieved. Fortunately, as discussed in Section 3.8, in many physical problems ω and $\sinh \chi$ depend on each other in a favorable manner. More precisely, the product $\omega \sinh \chi$ does not depend on frequency. This allows a reasonable coefficient (3.54) to be achieved at any evanescent frequency, resulting in a quite good absorption of evanescent waves in realistic problems. Such an achievement is illustrated in the next chapter for wave-structure interaction problems and waveguide problems.

CHAPTER 6

Optmization of the PML ABC in Wave-Structure Interaction and Waveguide Problems

In this chapter, use of the PML ABC in two typical applications of the FDTD method is described in detail. The numerical reflection observed from the PML is interpreted and the PML parameters are optimized so as to reduce the computational cost of the PML while preserving a satisfactory simulation of free space.

The considered applications are, first a typical wave-structure interaction problem like the ones faced in electromagnetic compatibility (EMC), second a waveguide problem.

6.1 WAVE-STRUCTURE INTERACTION PROBLEMS

Numerous applications of the FDTD method are wave-structure interaction problems, in the fields or Radar Cross Section (RCS), Bioelectromagnetics, or Electromagnetic Compatibility (EMC). In the latter case, typical problems are calculations of the fields and currents induced upon an object of interest by such incident waves as a lightning pulse or a nuclear electromagnetic pulse. These problems are open because the scattered field radiates toward the surrounding free space, so that an ABC is needed to limit the size of the computational domain.

To compute acceptable results in EMC problems, it is known that such ABCs as the one-way wave equation [9, 10] or the matched layer [13, 14] must be placed some distance from the scattering structure, out of the evanescent region. In general, the required distance is of the order of the largest size of the structure. With the PML ABC, simple numerical experiments easily show that the PML either can be placed quite close to the structure, or must be placed some distance away, depending on the PML parameters. In the latter case, the overall computational domain is as large as with previously used ABCs, while in the former case it is dramatically reduced. Thus, the design of the PML strongly impacts the overall size of the computational domain. In the following the optimization of PMLs to be placed in the evanescent region, close to the structure, is addressed. Two optimum PMLs are possible, the

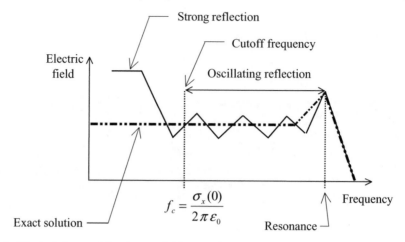

FIGURE 6.1: Typical shape of the electric field normal to the surface of a scattering structure, computed with a FDTD PML ABC placed close to the structure

first one based on use of the regular stretching factor (3.3), the second one based on the CFS factor (3.50).

6.1.1   The General Shape of the Results Computed with a PML Placed Close to a Structure

Various examples of wave-structure interaction results computed with PMLs close to the structure of interest can be found in [5], [41], [42]. If the PML is not properly designed, an important amount of spurious reflection occurs. This reflection cannot be explained with the continuous PML theory. It results from the FDTD discretization of space. The numerical reflection was analyzed empirically in [41] and interpreted theoretically later [46], [15]. In all the numerical experiments the electric field on the surface of scattering structures is correct for some time and then strongly departs from the exact solution. An example is shown in Fig. 5.2. In frequency domain, the electric field on the surface always is shaped as in Fig. 6.1. Below a certain frequency f_c depending on the conductivity implemented in the vacuum–PML interface $\sigma_x(0)$, a strong reflection is observed, i.e., the results strongly depart from the exact solution computed with a PML or any ABC set far away from the structure. From f_c to the resonance frequency of the structure, the results oscillate about the exact solution. And finally, above the resonance frequency no significant numerical reflection is present.

6.1.2   Interpretation of the Numerical Reflection

The shape of the spurious fields that contaminate the electric field on the surface of the structure (Fig. 6.1) can be interpreted easily by means of the theory of the numerical reflection derived in

Section 5.6. First, frequency f_c found on a pure empirical basis [41] is nothing but frequency (5.50) derived from the numerical theory. This means that the strong spurious field below f_c in Fig. 6.1 is due to the total reflection of evanescent waves whose $\cosh\chi$ is large enough so as condition (5.52) holds. In general, below f_c the field on the surface is several times larger than its exact value, because there are multiple reflections between the structure and the PML. Thus, when the regular stretching factor (3.3) is used, conductivity $\sigma_x(0)$ in the vacuum–PML interface is a critical parameter that plays a key role in the design of PMLs to be placed in the evanescent region.

To interpret the oscillatory reflection in Fig. 6.1, we must have an estimate of the parameter $\cosh\chi$ of the evanescent waves surrounding the scattering structure. Assuming that these waves decay as

$$e^{-\frac{\omega}{c}\sinh\chi\, d} \qquad\qquad (6.1)$$

where d is the distance from the structure, an estimate of $\sinh\chi$ can be found by observing that the characteristic length of the decay of evanescent waves around a structure is of the order of its largest size at any frequency lower than the resonance. For a structure of largest size w this means that (6.1) is small for $d = w$, so that we can write:

$$\frac{\omega}{c}\sinh\chi\, w = p \qquad\qquad (6.2)$$

where p is of the order of, or larger than, unity. For our purpose, the exact value of p is of little importance, what is important is the frequency dependence of $\cosh\chi$. From (6.2) we have

$$\cosh\chi = \sqrt{1 + \frac{p^2 c^2}{\omega^2 w^2}} \qquad\qquad (6.3)$$

which shows that $\cosh\chi$ is close to unity at the resonance frequency of the structure $\omega_0 = \pi c/w$. Finally, (6.3) can be rewritten as

$$\cosh\chi = \sqrt{1 + \frac{(\cosh^2\chi_0 - 1)f_0^2}{f^2}} \qquad\qquad (6.4)$$

where $\cosh\chi_0$ is the value of $\cosh\chi$ at the resonance f_0. Notice that $\cosh\chi$ and $\sinh\chi$ vary as $1/f$ for $f \ll f_0$, like in a waveguide for $f \ll f_{\text{cutoff}}$ (1.4b). Also, notice that $\omega\sinh\chi$ (6.2) does not depend on frequency, as a consequence of the fact that the characteristic length of the decay of evanescent waves is not frequency dependent.

Calculation of the numerical reflection from a PML with (5.57) and (6.4) clearly shows that the oscillatory reflection in Fig. 6.1 is due to the reflection of evanescent waves of the form (6.4). A detailed analysis can be found in [15]. At frequencies higher than f_c the evanescent

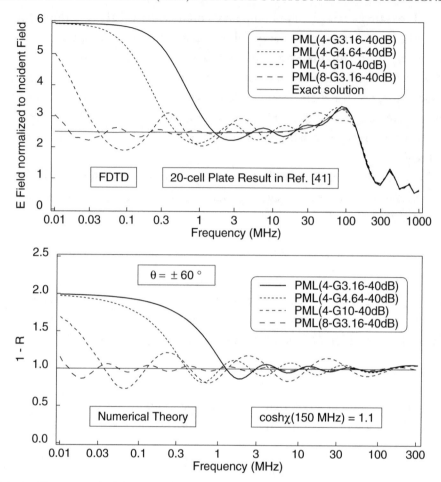

FIGURE 6.2: Comparison of the normal electric field computed by the FDTD method on a 20-cell-thin plate in [41] (upper part) with quantity $1 - R$ computed using (5.57) and (6.4) with $\cosh\chi_0 = 1.1$ (lower part)

waves penetrate within the PML with only a small reflection from the vacuum–PML interface. Then, in the PML where the conductivity grows from one FDTD cell to the next, these frequencies are partially reflected from the inner interfaces of the PML. As frequency grows the reflection is mainly due to an electric conductivity or a magnetic conductivity, i.e., the reflection is either positive or negative. From this, the field computed on the structure, which is the addition of the exact field with the reflected field, oscillates about the exact field in function of frequency, as represented in Fig. 6.1.

Fig. 6.2 shows an attempt to reconstructing a FDTD result in [41] by means of the numerical reflection (5.57) and assumption (6.4). The upper part from [41] gives the normal electric field on a 2D 20-cell-thin plate surrounded with various 4-cell PMLs set two cells from it. The lower part shows quantity $1 - R$ computed by (5.57) and (6.4) for the same four PMLs as in the FDTD calculation, at incidence $60°$. As observed, the oscillatory region closely resembles that of the FDTD result. The low frequency plateau is also like its FDTD counterpart, although its magnitude, which does not depend on θ ($R = -1$ in this region), is lower than the FDTD one. The difference is due to multiple reflections between the PML and the structure so that the field upon the structure is not simply $1 - R$. In conclusion, although such little arbitrary parameters as $\cosh \chi_0$ are of concern, the theory of numerical reflection allows all the characteristics of the FDTD results to be well reconstructed and interpreted in the evanescent region of the frequency spectrum, below the resonance frequency. This clearly demonstrates that the spurious reflection in interaction problems is due to evanescent fields with frequency dependence like (6.2).

6.1.3   Design of the PML Using a Regular Stretching Factor

In this paragraph, we consider PMLs with stretching factors (3.3). A practical method for designing a PML to be placed very close to the scattering structure, only two FDTD cells from it, has been proposed in [41] and [47]. The design of the PML is based on the existence of three critical parameters that impact the reflection from the PML ABC placed in evanescent fields.

The first critical parameter is the conductivity in the interface $\sigma_x(0)$ that must be small enough in order that all the frequencies of interest are above frequency f_c (5.50), because below f_c the reflection is total from the vacuum–PML interface. This can be written as

$$\sigma_x(0) = \frac{2\pi \varepsilon_0 f_{\min}}{\theta} \qquad (6.5)$$

where $f_{\min}$ is the smallest frequency of interest and θ is a margin factor that will not be larger than 10 in actual computations. For time domain calculations, $f_{\min}$ is about the inverse of the duration of the computation D_c, so that (6.5) can be rewritten as

$$\sigma_x(0) = \frac{2\pi \varepsilon_0}{\theta D_c}. \qquad (6.6)$$

The second critical parameter is the rate of increase of the conductivity in the PML, from $\sigma_x(0)$ in the vacuum–PML interface to its maximum value on the outer side of the PML. The magnitude of the numerical reflection in the oscillatory region in Fig. 6.1 depends on the ratio of successive conductivities. The larger this ratio, the larger the magnitude of the oscillatory reflection. This can be verified easily by means of FDTD experiments [41] and by means of numerical theory [15].

The third critical parameter is the theoretical reflection at normal incidence $R(0)$ which must be large enough so as to absorb the traveling waves. In EMC applications, moderate absorptions of the order of 40 dB are sufficient.

As a consequence of the three critical parameters, in a PML placed close to a scattering structure, in the evanescent region, the conductivity must grow from a small value given by (6.6) to a value large enough so as to render $R(0)$ equal to a prescribed value, while the step of increase of the conductivity from one FDTD cell to the next must remain smaller than another prescribed value. From these three constraints, the profile of conductivity that yields the thinner PML, in terms of FDTD cells, is the geometrical profile of conductivity (5.25). Conversely, with a polynomial profile the ratio of successive conductivities varies in the PML, leading to a thicker PML for a given level of residual reflection [41, 47].

The values to be set to the critical parameters in actual applications were estimated empirically [41, 47]. Fortunately, it was observed that they do not depend too much on the geometry and size of the scattering structure, with the exception of the ratio g of the geometrical profile of conductivity that depends on the size of the structure expressed in FDTD cells. Five consistent sets of critical parameters are given in [41, 47] in function of a synthetic parameter that expresses the expected accuracy of the results. Once the critical parameters θ, g, $R(0)$ are chosen, the PML thickness N, in cells, is given by the following formula obtained using (5.28a) and (6.6):

$$N = \frac{1}{\ln g} \ln \left[1 - \frac{c}{4\pi} (\sqrt{g} - 1) \ln R(0) \frac{\theta}{\Delta x} D_c \right]. \qquad (6.7)$$

With the critical parameters in [47], quite good results can be obtained on structures that are several hundreds of cells in length, with PMLs that are typically 10–15 cells in thickness. Two examples are shown in Figs. 6.3 and 6.4. The electric field at several points on a 500-cell-thin plate and on a 237-cell-long airplane is shown. The results were computed with the five sets of critical parameters corresponding to the five accuracies proposed in [47]. From the poorest accuracy to the highest one, the values of the synthetic parameter, denoted as p, read $-2, -1, 0,$ $+1, +2$, respectively. The PML–plate separation was only 2 FDTD cells. For every calculation the thickness of the PML and the three critical parameters (g, $R(0)$ in dB, and θ) are reported in the figures. With the best accuracy ($p = +2$) the results are superimposed to the exact solution computed with an ABC placed far from the structure (not shown in the figures). With the medium accuracy ($p = 0$) they are quite close to the exact solution. The corresponding PML thicknesses equal 15 cells with the plate, and 12 cells with the airplane. Even with the poorest accuracy ($p = -2$) the results can be viewed as acceptable in the context of EMC applications.

In summary, the critical parameters given in [47] permit reliable calculations to be performed with relatively thin PMLs placed only two FDTD cells from the scattering structure.

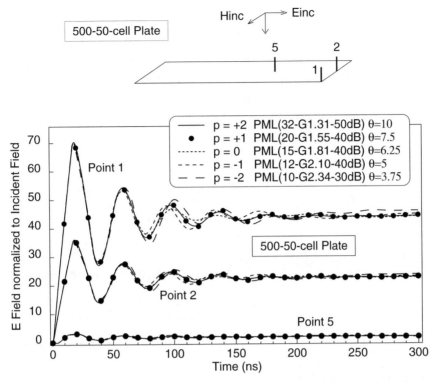

FIGURE 6.3: Electric field at three locations on a 500-50-cell plate stricken by an incident wave of the form $[1 - \exp(t/\tau)]$ with $\tau = 1$ ns. The space step is 5 cm. The PML ABC is placed 2 FDTD cells from the plate. Results are graphed for the five PMLs corresponding to the five consistent sets of PML parameters proposed in [47]. The $p = +2$ results are superimposed to the exact solution

In comparisons with previous ABCs that must be placed far from the structure [48], the computational times are typically reduced with a factor of 10 or more, while the results are better and more reliable.

Some refinements based on [23] allow the thickness of the above-optimized PML to be reduced by 2–3 FDTD cells [47]. They are not discussed here, because a new major improvement to the simulation of free space with a PML ABC is now at hand for solving wave-structure interaction problems [49]. This improvement consists of using the CFS stretching factor (3.50) in place of the regular factor (3.3).

6.1.4 Design of the PML Using the CFS Stretching Factor

With the stretching factor (3.50), the attenuation of evanescent waves (3.54) at frequencies lower than f_α (3.51) is widely smaller than with the regular stretching factor. Moreover, since around scattering structures $\omega \sinh\chi$ is constant from (6.2), in a CFS-PML placed close to a

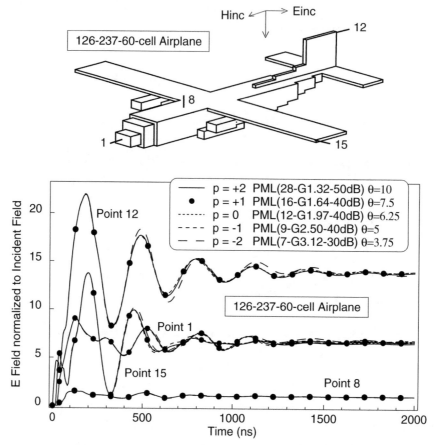

FIGURE 6.4: Electric field at four locations on a 126-237-60-cell airplane structure stricken by an incident wave of the form $[1 - \exp(t/\tau)]$ with $\tau = 2$ ns. The cell size is $25 \times 16.66 \times 16.66$ cm. The PML ABC is placed 2 FDTD cells from the airplane. Results are graphed for the five PMLs corresponding to the five consistent sets of PML parameters proposed in [47]. The $p = +2$ results are superimposed to the exact solution

structure the attenuation (3.54) does not depend on frequency below f_α. In consequence, by means of an adequate choice of f_α (3.51), or equivalently of α_x, the attenuation can be set equal to that of high frequency ($f \gg f_\alpha$) traveling waves that is like in a regular PML. This results in a reasonable attenuation of evanescent waves in the PML, neither too small nor too high, so as to remove the strong numerical reflection present when using the regular stretching factor.

To make equal the absorptions of evanescent waves (at $f \ll f_\alpha$) and traveling waves (at $f \gg f_\alpha$), let us equal the coefficient in (3.54) with the attenuation in a regular PML at normal incidence $\exp(-\sigma_x x/\varepsilon_0 c)$. This yields

$$\frac{f}{f_\alpha} \frac{\sigma_x}{\varepsilon_0 c} \sinh \chi \sin \theta \, x = -\frac{\sigma_x}{\varepsilon_0 c} x \tag{6.8}$$

where $\chi\theta < 0$ because it is assumed that the waves propagate and are evanescent toward $+x$. Using (3.51) and (6.2), the following α_x is obtained:

$$\alpha_x = \frac{c\varepsilon_0}{w} |p \sin \theta|. \tag{6.9}$$

Quantities p and $\sin\theta$ are unknown parameters of the evanescent waves that surround the scattering structure. What we know is that they are of the order of unity, so that we can say that the optimum α_x that permits the absorptions of evanescent and traveling waves to be equal is about:

$$\alpha_0 = \frac{c\varepsilon_0}{w}. \tag{6.10}$$

Frequency f_α corresponding to α_0 can be found from (3.51). Inserting the resonance frequency of the structure of size w, that is $f_0 = c/2w$, we obtain

$$f_{\alpha 0} = \frac{f_0}{\pi} \tag{6.11}$$

which shows that the resonance is close to frequency (3.51) if $\alpha_x = \alpha_0$. As in the case of the waveguide problem discussed in Chapter 3, the frequency of the transition between the two regimes of the CFS-PML is close to the frequency separating the evanescent waves with the traveling waves. Fig. 3.3 is valid for the wave-structure interaction problem, with only a correction with factor π from (6.11).

Fig. 6.5 shows a FDTD experiment, for the same 20-cell-thin plate problem as in Fig. 6.2. The upper part shows the E field at the end of the plate, for different values of α_x, from $\alpha_x = 0$ to $\alpha_x = 10\,\alpha_0$. The 4-cell-thick PML is placed 2 cells from the plate. For $\alpha_x = 0$ the result is close to that in Fig 6.2. As α_x grows the reflection at low frequency decreases. For $\alpha_x = \alpha_0$ the result is about undistinguishable from the exact solution. For $\alpha_x = 10\,\alpha_0$ the result departs from the reference, both at low frequency and around the resonance (150 MHz) where the attenuation is too small. The lower part of Fig. 6.5 shows an attempt to reconstructing the numerical reflection of the evanescent waves by means of the numerical theory (5.57) and assumption (6.2). As with the regular stretching factor in Fig. 6.2, the reconstruction is in a good agreement with the actual FDTD calculations. More details about the interpretation of the results in this example can be found in [25].

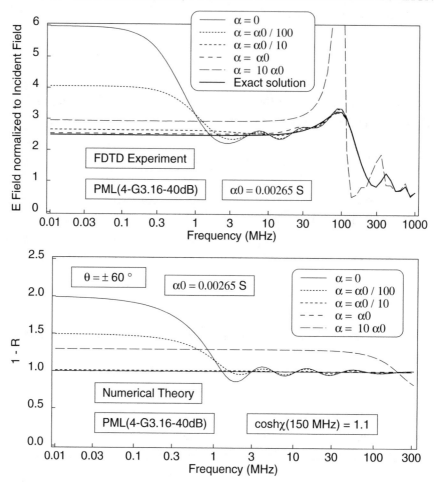

FIGURE 6.5: Comparison of the normal electric field computed using the FDTD method (upper part) with quantity $1 - R$ computed using (5.57) and (6.4) with $\cosh\chi_0 = 1.1$ (lower part), for a CFS-PML ABC and a 20-cell-thin plate

Fig. 6.6 shows a 3D numerical experiment with a 500-cell-thin plate stricken with a unit-step incident wave. As in the 2D case of Fig. 6.5, the improvement is dramatic with α_x value (6.10). For $\alpha_x < \alpha_0$ a strong reflection of low frequency evanescent waves is observed. For $\alpha_x > \alpha_0$ the absorption around the resonance frequency of the plate is too small (oscillations in the result). This confirms that the best value of α_x is close to (6.10).

By increasing the thickness of the CFS-PML, which equals four FDTD cells in Fig. 6.6, the results for $\alpha_x = \alpha_0$ become closer to the exact solution. With a 6-cell CFS-PML the results are as good as with the 10-cell regular PML in Fig. 6.3. One may think that the accuracy could be improved at will by increasing the thickness of the CFS-PML. That is not true. Actually, as

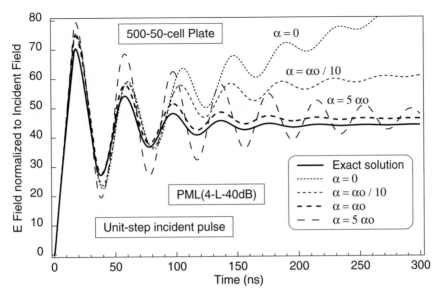

FIGURE 6.6: *E* field on a 500-50-cell plate computed with a CFS-PML ABC placed 2 cells from it, with various values of the CFS-PML parameter α

the thickness grows the computed results tend to a limit which slightly differs from the exact solution. This is because the absorption of traveling waves is smaller than the prescribed $R(0)$ at frequencies around transition $f_{\alpha 0}$.

In order to improve and control the accuracy of the results computed with $\alpha_x = \alpha_0$, various solutions were envisaged and tested [49]. These investigations have shown that:

- using a parameter α_x decreasing from the vacuum–PML interface to the outer boundary of the PML is an improvement. Doing this, the transition frequency f_α decreases from a value $f_{\alpha 1} > f_{\alpha 0}$ in the interface to a value $f_{\alpha 2} < f_{\alpha 0}$ on the outer boundary. This improves the attenuation in the critical region of the transition between evanescent and traveling waves, close to $f_{\alpha 0}$.

- the ratio $\sigma_x(0)/\alpha_x(0)$ is a critical parameter, where $\sigma_x(0)$ and $\alpha_x(0)$ are σ_x and α_x implemented in the vacuum–CFS–PML interface. The following must hold:

$$\frac{\sigma_x(0)}{\alpha_x(0)} \leq 1 \qquad (6.12)$$

From this, an optimum CFS-PML has been proposed as follows [49]. In the PML, from the interface to the outer boundary, α_x decreases geometrically from $50\alpha_0$ to $\alpha_0/5$, and the

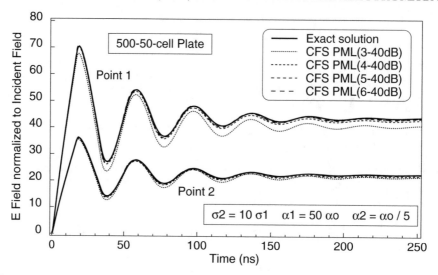

FIGURE 6.7: *E* field on a 500-50-cell plate computed with the optimized CFS PML. The conductivity grows geometrically in the PML, from σ_1 to $10\sigma_1$, and α decreases geometrically from $50\alpha_0$ to $\alpha_0/5$. Results are graphed for PMLs 3, 4, 5, 6 cells in thickness, set 2 cells from the plate, with $R(0) = -40$ dB

conductivity grows geometrically by a factor of 10, from $\sigma_x(0)$ to $10\sigma_x(0)$. For a PML of thickness N, the ratio g is given by $g = 10^{1/N}$, and the conductivity is set by means of (5.27) in order that the normal reflection equals a prescribed value $R(0) = -40$ dB.

Results computed with such profiles of α_x and σ_x are shown in Fig. 6.7 for the 500-cell plate, with various thicknesses of the PML from 3 FDTD cells to 6 FDTD cells. As observed, with the 5-cell PML the results are about superimposed to the reference solution. With the 4-cell PML the results are as good as those computed with the optimum 10-cell regular PML in Fig. 6.3. The performance of the CFS-PML is widely better than that of the regular PML for this example. This also holds with other structures, as shown in Fig. 6.8 for the same airplane as in Fig. 6.4. The results are quite good with a CFS-PML only 3 cells in thickness, and are perfect (superimposed to the exact solution) with a 4-cell CFS-PML. With other incident waves like nuclear pulses or lightning pulses whose low frequency content is poorer, the results are better than with the unit-step wave. Several tests were performed with plates of various lengths from 100 cells to 1000 cells. The required thickness of the CFS-PML, in FDTD cells, does not depend too much on this length, so that the results in Figs. 6.7 and 6.8 can be viewed as representative of the worst case for scattering structures than can be handled by current computers, that is for structures 200–1000 FDTD cells in size. From this, the con-clusion is that a CFS-PML placed two FDTD cells from the structure, with conductivity,

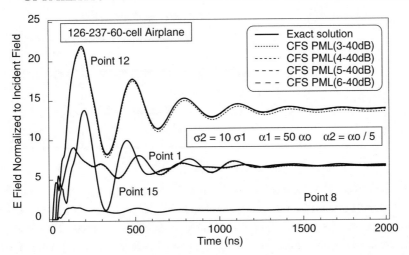

FIGURE 6.8: *E* field on a 126-237-60 airplane computed with the optimized CFS PML. The computational conditions are like in Fig. 6.7

α_x parameter, and $R(0)$ prescribed in the above, allows the following solutions to be computed:

- reference solutions with a 6-cell CFS-PML.

- accurate solutions with a 4-cell CFS-PML.

- acceptable solutions with a 3-cell CFS-PML.

This is significantly better than with the optimized PML based on the regular stretching factor. The thickness of the PML is reduced with a factor larger than 2. As a conclusion, in wave-structure interaction problems, the optimized CFS-PML placed only two cells from the scattering structure allows the cost of the simulation of free space to be negligible in comparison with the cost of the region of interest of the computational domain.

6.2   WAVEGUIDE PROBLEMS

Modes in waveguides are evanescent in the longitudinal direction below the cutoff frequency. For a 2D parallel plate, the waveform and the evanescence parameter χ are given by (1.4). The phase propagates in the transverse direction y and the magnitude decreases in the longitudinal direction x, according to (1.4a). Consider a PML perpendicular to x with the stretching factor (2.6). The waveform in the PML is given by (2.30) with $\theta = \pm\pi/2$ and $\theta\chi < 0$. For instance, in the case $\theta = \pi/2$:

$$\psi = \psi_0 \, e^{j\omega\left[t - \frac{\cosh\chi}{c}y - \frac{\sigma_x}{\varepsilon_0 c\omega}\sinh\chi x\right]} e^{\frac{\omega}{c}\sinh\chi x}. \tag{6.13}$$

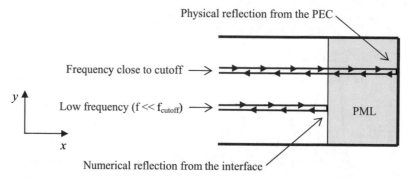

FIGURE 6.9: Reflection of evanescent waves from a PML ending a waveguide. The phase varies in the transverse direction y and the amplitude is evanescent in the longitudinal direction x. The lines with arrows symbolize planes of constant phase of the incident and reflected waves. The arrows are oriented toward the decrease of the magnitude, to the right for the incident waves, to the left for the reflected waves

The real exponential in (6.13) is the natural decrease of an evanescent wave in a vacuum. There is no additional attenuation in the PML, because the propagation of the phase is perpendicular to the vacuum–PML interface ($\cos \theta = 0$). Nevertheless, the waveform in the PML differs from the waveform in a vacuum (1.4) due to the presence of an additional term in the phase. This term grows as the frequency decreases. If the frequency is small enough, the phase varies rapidly with distance. This may result in a strong numerical reflection from the PML if the variation is rapid in comparison with the size of the FDTD cell, because then the discretized space cannot sample accurately the field. This phenomenon was first analyzed in [50]. Finally, two spurious reflections occur as using a PML ABC to absorb the evanescent waves at the end of a waveguide (Figs. 6.9 and 6.10). First the PML does not absorb these waves, in theory, so that they are reflected from the PEC that ends the PML. This mainly results in the reflection of frequencies that are close to the cutoff, because their natural decrease is weak. Second, at low frequency a strong numerical reflection occurs from the vacuum–PML interface, due to the phase term in (6.13). This has been illustrated clearly in [50], where FDTD experiments and results of calculations with the FDTD numerical theory have been reported. The reflection from the interface is also clearly visible in Fig. 6.10, where the low frequency reflection is larger than the natural decay of the waves.

6.2.1   Improvement of the Absorption by Means of a Real Stretch of Coordinates

In order to increase the decay of evanescent waves in the PML, a real stretch was introduced [32] in the stretching factor (2.6), like the parameter κ in the CFS factor (3.50). This extends the physical thickness of the PML so that the natural decay of the waves is increased. A factor

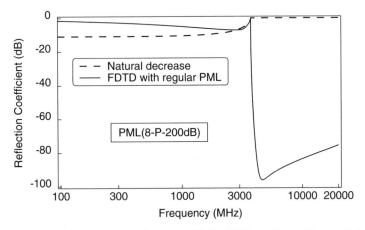

FIGURE 6.10: Reflection of the TM_1 mode from a PML ABC ending a 2D parallel-plate waveguide. The cutoff frequency equals 3.75 GHz (waveguide width 40 mm). The FDTD results are computed with a regular stretching factor (2.6) in the PML. The natural decrease is for a range equal to two PML thicknesses

κ growing from the vacuum–PML interface to the outer boundary of the PML is used, but the maximum value of the stretch is limited because large FDTD cells result in numerical reflection of the highest frequencies. Moreover, the method does not reduce the strong reflection due to the phase term in (6.2). Nevertheless, this simple modification of the PML results in a reduction of the overall reflection, but limited to a band of frequency below the cutoff, as illustrated in [32].

6.2.2   Improvement of the Absorption by Using a CFS Stretching Factor

A far better method to absorb effectively the evanescent waves at the end of a waveguide [51] consists of using the CFS stretching factor (3.50). Above frequency f_α (3.51) the CFS-PML is like a regular PML, but below f_α the CFS-PML reduces to a real stretch of coordinates and the waveform becomes (3.53), (3.54). In the case of a waveguide the evanescent waves $(\theta = \pi/2)$ are absorbed in the CFS-PML, due to the real exponential in (3.53), (3.54), and the phase term that tends to infinity at low frequency in (6.13) is removed. From this, the two kinds of spurious reflections present with a regular PML (Fig. 6.9) are removed below f_α.

As discussed at the end of Chapter 3, because the product $\omega \sinh\chi$ is constant far below the cutoff frequency of the considered mode, the absorption of evanescent waves in (3.54) is not frequency dependent. This absorption can be set equal to that of traveling waves at $f \gg f_\alpha$ by choosing α_x equal to α_0 (3.58). Then f_α coincides with the cutoff frequency (3.59) so that the CFS-PML is very well suited to the absorption of the entire spectrum of waves present in the waveguide (Fig. 3.3).

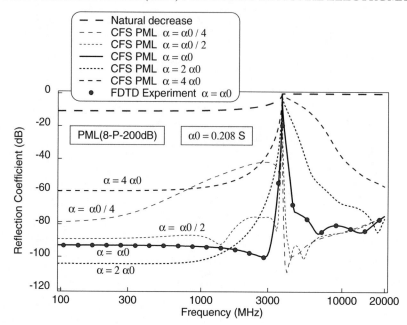

FIGURE 6.11: Reflection of the TM_1 mode from a PML ABC ending the same parallel-plate waveguide as in Fig. 6.10. The stretching factor is the CFS factor (3.50). Various parameters α are considered around the optimum value α_0 (3.58). The reflections have been computed theoretically using (5.57) and (1.4), with in addition a FDTD simulation for the case $\alpha = \alpha_0$. Notice the perfect agreement of the simulation with (5.57)

Results in Fig. 6.11 show the effect of the value of α_x on the reflection from a CFS-PML. With $\alpha_x = \alpha_0$ the improvement is dramatic in comparison with the regular PML (Fig. 6.10), the reflection drops below -80 dB with the 8-cell-thick CFS-PML used in the experiment. Only a narrow band of frequencies is reflected around the cutoff frequency. This is because $\sinh \chi$ vanishes as frequency tends to the cutoff, so that the absorbing term in (3.54) tends to unity. For α_x larger than α_0 the reflection grows because the attenuation in (3.54) is too small (f_α too large in the exponential).

From other experiments reported in [51] the reflection is not too much sensitive to the design of the PML. A quite small reflection in the evanescent region (-80 dB or less) is obtained with a polynomial profile of conductivity of moderate power (2 or 3), a theoretical reflection $R(0)$ of the order of -200 dB, and a PML thickness in 6–10 cells.

In the above, it is assumed implicitly that only one mode is present in the waveguide, because the optimum α_x (3.58) depends on the order n of the mode. In the case where several modes are present, α_x cannot be optimum for all the modes. This has never been experienced in the literature. A way to solve this problem is suggested by Fig. 6.11 where the reflection in the evanescent region is about -60 dB from $\alpha_x = \alpha_0/4$ to $4\alpha_0$. This suggests that the reflection is not very sensitive to the value of α_x in the vicinity of α_0. From this, when several modes are present, a good compromise may be choosing α_x as the geometric average value of α_0 of the lowest cutoff frequency with α_0 of the highest cutoff frequency. As long as the ratio of these two cutoff frequencies is not too large, all the modes will be absorbed with a small reflection (about -60 dB in the case of Fig. 6.11 if the ratio of cutoffs is not larger than 16). Another solution that could be envisaged if several modes are present may be use of a parameter α_x decreasing in the PML, like that used in the optimized CFS-PML for wave-structure interaction problems in the previous section.

6.3   CONCLUDING REMARKS TO THE APPLICATION OF THE PML ABC TO FDTD PROBLEMS

In the above the optimization of the PML ABC has been addressed in two cases that are important applications of the FDTD method. The optimized PMLs do absorb the evanescent waves so that they can be placed in the evanescent regions, that is as close as possible to the region of interest, namely the scattering structure or the end of the waveguide. The PML performance is then far better than that of previously used ABCs that reflect the evanescent waves.

What has to be noticed is that optimizing a PML for a given application of the FDTD method requires a good understanding of the propagation and absorption of waves in the PML, especially of evanescent waves, both in the continuous theoretical space and in the FDTD discretized space. Another required knowledge is the general form of the waves to be absorbed. All this permits the origin of the numerical reflection from FDTD PMLs to be analyzed, and then an optimum PML to be designed accordingly.

As shown with two problems in the above, nowadays the CFS-PML based on the stretching factor (3.50) seems the best way in view of optimizing PMLs in realistic problems. This is because in most physical applications, at least as long as wide band or time domain calculations are of concern, both evanescent and traveling waves are involved. As long as the evanescence coefficient varies as the inverse of frequency, as in waveguides and around scattering structures, the CFS-PML permits the absorption to be reasonable and about constant in the whole spectrum, resulting in a small numerical reflection at any frequency, even with relatively thin PMLs placed in the close vicinity of the region of interest. The CFS-PML is a naturally optimized PML well suited to the physics in waveguide and wave-structure interaction problems. This is also the case in other problems, for instance in the scattering by periodic corrugated surfaces where the scattered wave can be expanded in Floquet modes [52] that are like (1.4a), with (1.4b) replaced with

$$\sinh \chi = \pm \sqrt{\left(\sin \theta + \frac{n\pi c}{b\omega} \right)^2 - 1} \tag{6.14}$$

where θ is the incidence of the incident wave and b is the half-period. From (6.14), $\omega \sinh\chi$ is constant at low frequency. This leads to the optimum value of α_x given by the same formula as in the waveguide case (3.58), with half-period b in place of width a.

CHAPTER 7

Some Extensions of the PML ABC

In the previous chapters the PML ABC has been derived in Cartesian coordinates for solving the Maxwell equations, and the only numerical method that has been considered in detail is the Yee FDTD method. In the following, some extensions of the PML concept to other coordinates and other numerical techniques are briefly reviewed.

7.1 THE PERFECTLY MATCHED LAYER IN OTHER SYSTEMS OF COORDINATES

Several papers can be found in the literature about the extension of the PML medium to various systems of coordinates, mainly in the context of the FDTD method. In [53–60] PML ABCs are derived for cylindrical and spherical coordinates, generalized curvilinear coordinates, and nonorthogonal coordinates. In [61], the PML ABC is extended to the body of revolution (BOR) FDTD method. In most cases the PML is derived by means of a stretch of coordinates in the direction normal to the interface.

Paper [56] discusses an important issue on the PML ABC. It is shown by means of theoretical derivations that the FDTD-PML in cylindrical and spherical coordinates must be concave. The stability is not ensured with convex PMLs. This is in accordance with the intuitive feeling that the PML can absorb radiated fields but cannot replace sources. Each point in a volume space bounded with a PML must be in direct visibility from any other point of this volume, i.e., the interface between the inner medium and the PML must be either plane, that is the Cartesian case, or concave, in the general case. This is a limitation to the use of the PML ABC in some special problems addressed in [8].

7.2 THE PERFECTLY MATCHED LAYER WITH OTHER NUMERICAL TECHNIQUES

Introduced in the context of the FDTD method, the PML ABC has been rapidly extended to other techniques used in numerical electromagnetics. To do this, the interpretation in terms of stretched coordinates [16, 17] has been very fruitful.

First, PML ABCs have been derived for other finite-difference schemes than the Yee second-order scheme addressed in Chapters 5 and 6. A PML ABC for the four-order scheme is described in [62]. In [39] the PML is used with the pseudospectral time-domain (PSTD) method. The PML ABC is also used with the finite-volume time-domain (FVTD) method [63] and with the finite-difference frequency-domain (FDFD) method [17, 64]. It is also of current use with the unconditionally stable ADI-FDTD method [65, 66].

In the Transmission Line Matrix (TLM) method, the PML ABC has been implemented in two ways: first by surrounding the TLM computational domain with a layer of FDTD cells where the PML is placed [67], second by implementing a true TLM-PML, that is a PML medium discretized with the TLM method [68, 69].

One of the early extensions of the PML concept was the introduction of the uniaxial PML (3.48), (3.49) for use as an ABC in the frequency-domain finite-element method [20]. As discussed in Chapter 3 the uniaxial PML corresponds to a stretch of fields in place of the stretch of coordinates of the regular PML. Conversely to the regular PML, the uniaxial PML is Maxwellian, i.e., it is governed by the regular Maxwell equations of an anisotropic medium with permittivity and permeability tensors (3.49). From this it can be implemented in the FEM method in a natural way [21]. The PML ABC can also be used with the time domain finite element method [70].

Use of the PML ABC has been reported with the paraxial (parabolic) equation method. Initially developed in the context of seismic [71], it is also used in numerical electromagnetics [72]. The PML ABC has also been implemented in the Beam Propagation method based on the finite difference solution of the Helmoltz equation [73].

The PML ABC has been used for nonlinear calculations in the case of the propagation of solitons, with the FDTD method [74, 75]. In that case, a special difficulty lies in the fact that the permittivity is field dependent in the medium, so that the matching condition (1.9) to be satisfied by the PML conductivities is also field dependent. At least in the cases reported in [74, 75], this problem has been overcome easily by using a simple iterative procedure.

Finally, in the domain of electromagnetics, use of the PML ABC has been reported in particle in cell (PIC) finite difference calculations where moving charged particles are taken into account [76].

In all the numerical methods, the PML ABC widely improves the simulation of free space in comparison with previously used ABCs. Nevertheless, the PML ABC is never perfect in the discretized space, especially when evanescent waves are of concern. No analysis of the numerical reflection like that in [15] has been published in the literature for other methods than the FDTD one. But we can predict that what is observed with the FDTD method is also valid with the other techniques, i.e., when the waves are so strongly evanescent that they must be absorbed upon less than one cell, one volume, or one element, as schematized in Fig. 5.6, the

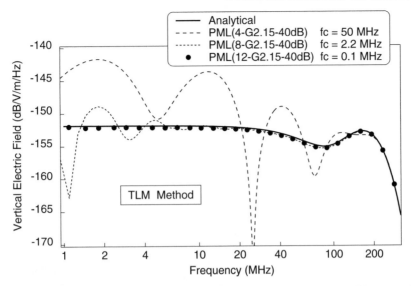

FIGURE 7.1: Electric field radiated from a short dipole, computed with the TLM method in [69]. The computational setup is the same as in Fig. 5.3

numerical scheme cannot sample properly the wave, resulting in a strong numerical reflection. The only numerical experiment performed to assess this expectation is that reported in [69]. This test consisted of computing the field radiated from a dipole using the TLM method. The results are shown in Fig. 7.1. The computational conditions were exactly the same as in the FDTD dipole experiment in [42] whose results are reproduced in Fig. 5.3. As observed, the results in Fig. 7.1 look like the ones in Fig. 5.3. Especially, with both methods the results are correct at high frequency and then depart from the exact solution below frequency f_c (5.50) that equals 50.3 MHz and 2.25 MHz for the 4-cell and 8-cell PMLs. This means that the strongly evanescent waves are reflected from the vacuum–PML interface as in the case of the FDTD method. From this, what is stated in Chapter 6 about the choice and optimization of FDTD PMLs could also be applied to the choice and optimization of the TLM-PML. That is probably true with other numerical techniques, although no explicit study of the numerical reflection has been reported in the literature.

7.3   USE OF THE PERFECTLY MATCHED LAYER WITH OTHER EQUATIONS OF PHYSICS

By means of a complex stretch of coordinates, the PML ABC has been extended to most partial differential equations that govern other domains of physics where unbounded problems have to be solved. In the literature, a number of papers can be found on use of the PML ABC, such as [77, 78] in acoustics, [79] in elastodynamics, [71] in seismic, or [80, 81] in hydrodynamics.

Bibliography

[1] R. F. Harrington, *Field Computation by Moment Methods*. New York: MacMillan, 1968.

[2] K. S. Yee, "Numerical solution of initial boundary value problems involving Maxwell's equations in isotropic media," *IEEE Trans. Antennas Propagat.*, vol. 14, pp. 302–307, 1966. doi:10.1109/TAP.1966.1138693

[3] A. Taflove and M. E. Brodwin, "Numerical solution of steady-state electromagnetic scattering problems using the time-dependant Maxwell's equations," *IEEE Trans. Microw. Theory Tech.*, vol. 23, pp. 623–630, 1975. doi:10.1109/TMTT.1975.1128640

[4] R. Holland, "THREDE: A free-field EMP coupling and scattering code," *IEEE Trans. Nucl. Sci.*, vol. 24, pp. 2416–2421, 1977.

[5] J.-P. Bérenger, "A perfectly matched layer for the absorption of electromagnetic waves," *J. Comput. Phys.*, vol. 114, pp. 185–200, 1994. doi:10.1006/jcph.1994.1159

[6] O. M. Ramahi, "Complementary operators: A method to annihilate artificial reflections arising from the truncation of the computational domain in the solution of partial differential equations," *IEEE Trans. Antennas Propagat.*, vol. 43, pp. 697–704, 1995. doi:10.1109/8.391141

[7] M. J. Grote and J. B. Keller, "Non reflecting boundary conditions for Maxwell's equations," *J. Comput. Phys.*, vol. 139(2), pp. 327–342, 1988.

[8] B. Shanker, M. Lu, A. Arif Ergin, and E. Michielssen, "Plane-wave time-domain radiation boundary kernels for FDTD analysis of 3D electromagnetic phenomena," *IEEE Trans. Antennas Propagat.*, vol. 53, pp. 3704–3716, 2005. doi:10.1109/TAP.2005.858590

[9] B. Engquist and A. Majda, "Absorbing boundary conditions for the numerical simulation of waves," *Math. Comput.*, vol. 31, pp. 629–651, 1977. doi:10.2307/2005997

[10] G. Mur, "Absorbing boundary conditions for the finite-difference approximation of the time-domain electromagnetic field equations," *IEEE Trans. Electrom. Compat.*, vol. 23, pp. 377–382, 1981.

[11] L. N. Trefethen and L. Halpern, "Well-posedness of one-way wave equations and absorbing boundary conditions," *Math. Comput.*, vol. 47, pp. 421–435, 1986. doi:10.2307/2008165

[12] R. Higdon, "Numerical absorbing boundary conditions for the wave equations," *Math. Comput.*, vol. 49, pp. 65–90, 1987. doi:org/10.2307/2008250

[13] R. Holland and J. Williams, "Total-field versus scattered-field finite-difference: A comparative assessment," *IEEE Trans. Nucl. Sci.*, vol. 30, pp. 4583–4588, 1983.

[14] J.-P. Bérenger, "Calcul de la diffraction à l'aide d'une méthode aux différences finies," *CEM-83 Proc., Trégastel, France*, June 1983.

[15] J.-P. Bérenger, "Evanescent waves in PML's: Origin of the numerical reflection in wave-structure interaction problems," *IEEE Trans. Antennas Propagat.*, vol. 47, pp. 1497–1503, 1999. doi:10.1109/8.805891

[16] W. C. Chew and W. H. Weedon, "A 3D perfectly matched medium from modified Maxwell's equations with stretched coordinates," *Microw. Opt. Technol. Lett.*, vol. 7–13, pp. 599–604, 1994.

[17] C. M. Rappaport, "Perfectly matched absorbing conditions based on anisotropic lossy mapping of space," *IEEE Microw. Guid. Wave Lett.*, vol. 5–3, pp. 90–92, 1995. doi:10.1109/75.366463

[18] R. Mittra and U. Pekel, "A new look at the perfectly matched layer (PML) concept for the reflectionless absorption of electromagnetic waves," *IEEE Microw. Guid. Wave Lett.*, vol. 5–3, pp. 84–86, 1995. doi:10.1109/75.366461

[19] R. Mittra and U. Pekel, "An efficient implementation of Berenger's perfectly matched layer (PML) for finite difference time domain mesh truncation," *IEEE Microw. Guid. Wave Lett.*, vol. 6–2, pp. 94–96, 1996. doi:10.1109/75.482000

[20] Z. S. Sacks, D. M. Kingsland, R. Lee, and J.-F. Lee, "A perfectly matched anisotropic absorber for use as an absorbing boundary condition," *IEEE Trans. Antennas Propagat.*, vol. 43, pp. 1460–1463, 1995. doi:10.1109/8.477075

[21] J. L. Volakis, A. Chatterjee, and L. C. Kempel, *Finite-Element Method for Electromagnetics*. Piscataway, NJ: IEEE Press, 1998.

[22] J. Fang and Z. Wu, "Generalized perfectly matched layer—An extension of Berenger's perfectly matched layer boundary condition," *IEEE Microw. Guid. Wave Lett.*, vol. 5–12, pp. 451–453, 1995. doi:10.1109/75.481858

[23] J.-P. Bérenger, "Improved PML for the FDTD solution of wave-structure interaction problems," *IEEE Trans. Antennas Propagat.*, vol. 45, pp. 466–473, 1997. doi:10.1109/8.558661

[24] M. Kuzuoglu and R. Mittra, "Frequency dependence of the constitutive parameters of causal perfectly matched absorbers," *IEEE Microw. Guid. Wave Lett.*, vol. 6, pp. 447–449, 1996. doi:10.1109/75.544545

[25] J.-P. Bérenger, "Numerical reflection from FDTD-PML's: A comparison of the split PML with the unsplit and CFS PML's," *IEEE Trans. Antennas Propagat.*, vol. 50, pp. 258–265, 2002. doi:10.1109/8.999615

[26] T.-B. Yu, B. h. Zhou, and B. Chen, "An unsplit formulation of the Berenger's PML absorbing boundary condition for FDTD meshes," *IEEE Microw. Wirel. Comp. Lett.*, vol. 13, pp. 348–350, 2003. doi:10.1109/LMWC.2003.815694

[27]   J. A. Roden and S. D. Gedney, "Convolutional PML (CPML): An efficient FDTD implementation of the CFS-PML for arbitrary media," *Microw. Opt. Technol. Lett.*, vol. 27, pp. 334–339, Dec. 2000. doi:10.1002/1098-2760(20001205)27:5<334::AID-MOP14>3.0.CO;2-A

[28]   S. A. Cummer, "A simple, nearly perfectly matched layer for general electromagnetic media," *IEEE Microw. Wirel. Lett.*, vol. 13, no. 3, pp. 128–130, 2003. doi:10.1109/LMWC.2003.810124

[29]   J.-P. Bérenger, "On the reflection from Cummer's nearly perfectly matched layer," *IEEE Microw. Wirel. Lett.*, vol. 14, no. 7, pp. 334–336, 2004. doi:10.1109/LMWC.2004.829272

[30]   W. Hu and A. Cummer, "The nearly perfectly matched layer is a perfectly matched layer," *Antennas Wirel. Propagat. Lett.*, vol. 3, pp. 137–140, 2004.doi:10.1109/LAWP.2004.831077

[31]   S. D. Gedney, "An anisotropic perfectly matched layer-absorbing medium for the truncation of FDTD lattices," *IEEE Trans. Antennas Propagat.*, vol. 44, pp. 1630–1639, 1996. doi:10.1109/8.546249

[32]   J. Fang and Z. Wu, "Generalized perfectly matched layer for the absorption of propagating and evanescent waves in lossless and lossy media," *IEEE Trans. Microw. Theory Tech.*, vol. 44, pp. 2216–2222, 1996.

[33]   S. D. Gedney, "An anisotropic PML absorbing medium for the FDTD simulation of fields in lossy and dispersive media," *Electromagnetics*, vol. 16, pp. 399–415, 1996.

[34]   S. Gonzales Garcia, I. Villo-Perez, R. Gomez Martin, and B. Garcia Almedo, "Applicability pf the PML absorbing boundary condition to dielectric anisotropic media," *Electron. Lett.*, vol. 32, pp. 1270–1271, 1996. doi:10.1049/el:19960844

[35]   I. Villo-Perez, S. Gonzales Garcia, R. Gomez Martin, and B. Garcia Almedo, "Extension of Berenger's absorbing boundary condition to match dielectric anisotropic media," *IEEE Microw. Guid. Wave Lett.*, vol. 7, pp. 302–304, 1997. doi:10.1109/75.622549

[36]   A. P. Zhao, J. Juntunen, and A. V. Raisanen, "Material independent PML absorbers for arbitrary anisotropic dielectric media," *Electron. Lett.*, vol. 33, pp. 1535–1536, 1997. doi:10.1049/el:19971050

[37]   F. L. Teixeira and W. C. Chew, "General closed-form PML conductive tensors to match arbitrary bianisotropic and dispersive media," *IEEE Microw. Guid. Wave Lett.*, vol. 8, pp. 223–225, 1998. doi:10.1109/75.678571

[38]   F. L. Teixeira and W. C. Chew, "A general approach to extend Berenger's absorbing boundary condition to anisotropic and dispersive media," *IEEE Trans. Antennas Propagat.*, vol. 46, pp. 1386–1387, 1998. doi:10.1109/8.719984

[39]   Q. H. Liu, "PML and PSTD algorithm for arbitrary lossy anisotropic media," *IEEE Microw. Guid. Wave Lett.*, vol. 9, pp. 48–50, 1999. doi:10.1109/75.755040

[40] A. Taflove and S. C. Hagness, *Computational Electrodynamics. The Finite-Difference Time-Domain Method*. Boston, MA: Artech House, 2000.

[41] J.-P. Bérenger, "Perfectly matched layer for the FDTD solution of wave-structure interaction problems," *IEEE Trans. Antennas Propagat.*, vol. 44, pp. 110–117, 1996. doi:10.1109/8.477535

[42] J.-P. Bérenger, "Three-dimensional perfectly matched layer for the absorption of electromagnetic waves," *J. Comput. Phys.*, vol. 127, pp. 363–379, 1996. doi:10.1006/jcph.1996.0181

[43] W. C. Chew and J. M. Jin, "Perfectly matched layer in the discretized space: an analysis and optimization," *Electromagnetics*, vol. 16, pp. 325–340, 1996.

[44] J. Fang and Z. Wu, "Closed-form expression of numerical reflection coefficient at PML interfaces and optimization of PML performance," *IEEE Microw. Guid. Wave Lett.*, vol. 6, pp. 332–334, 1996. doi:10.1109/75.535836

[45] M.-S. Tong, Y. Chen, M. Kuzuoglu, and R. Mittra, "A new anisotropic perfectly matched layer medium for mesh truncation in finite difference time domain analysis," *Int. J. Electron.*, vol. 86, pp. 1085–1091, 1999. doi:10.1080/002072199132860

[46] J.-P. Bérenger "Numerical reflection of evanescent waves from perfectly matched layers," *Proc. IEEE Antennas Propagat. Symp., Montreal, Canada*, pp. 1888–1891, July 1997.

[47] J.-P. Bérenger, "Making use of the PML absorbing boundary condition in coupling and scattering FDTD computer codes," *IEEE Trans. Electrom. Compat.*, vol. 45, pp. 189–197, 2003. doi:10.1109/TEMC.2003.810803

[48] J.-P. Bérenger, "A perfectly matched layer for free-space simulation in finite-difference computer codes," *Ann. Télécommun.*, vol. 51, no. 1–2, pp. 39–46, Jan. 1996.

[49] J.-P. Bérenger, "Application of the CFS PML to the FDTD solution of wave-structure interaction problems," *Proc. IEEE Antennas Propagt. Symp, Columbus, USA*, pp. 984–987, June 2003. doi:10.1109/75.465042

[50] J. De Moerloose and M. A. Stuchly, "Behaviour of Berenger's ABC for evanescent waves," *IEEE Microw. Guid. Wave Lett.*, vol. 5, pp. 344–346, 1995.

[51] J.-P. Bérenger, "Application of the CFS PML to the absorption of evanescent waves in waveguides," *IEEE Microw. Wirel. Comp. Lett.*, vol. 6, pp. 218–220, 2002.

[52] J. A. Kong, *Electromagnetics Waves*. New York: John Wiley & sons, 1974.

[53] F. L. Teixeira and W. C. Chew, "Systematic derivation of anisotropic PML absorbing media in cylindrical and spherical coordinates," *IEEE Microw. Guid. Wave Lett.*, vol. 7, pp. 371–373, 1997. doi:10.1109/75.641424

[54] B. Yang and P. G. Petropoulos, "Plane-wave analysis and comparison of split-field, biaxial, and uniaxial PML methods as ABCs for pseudospectral electromagnetic wave

simulations in curvilinear coordinates," *J. Comput. Phys.*, vol. 146, pp. 747–774, 1998. doi:10.1006/jcph.1998.6082

[55] J. A. Roden and S. D. Gedney, "Efficient implementation of the uniaxial-based PML media in three-dimensional nonorthogonal coordinates with the use of the FDTD technique," *Microw. Opt. Technol. Lett.*, vol. 14, pp. 71–75, 1997. doi:10.1002/(SICI)1098-2760(19970205)14:2<71::AID-MOP1>3.0.CO;2-I

[56] F. L. Teixeira and W. C. Chew, "On causality and dynamic stability of perfectly matched layers for FDTD simulation," *IEEE Trans. Antennas Propagat.*, vol. 47, pp. 775–785, 1999. doi:10.1109/8.774130

[57] E. A. Navarro, C. Wu, P. Y. Chung, and J. Litva, "Application of PML superabsorbing boundary condition to non-orthogonal FDTD method," *Electron. Lett.*, vol. 30, pp. 1664–1666, 1994. doi:10.1049/el:19941120

[58] C. Wu , E. A. Navarro, P. Y. Chung, and J. Litva, "Modeling of waveguide structures using the nonorthogonal FDTD method with a PML absorbing boundary," *IEEE Microw. Opt. Technol. Lett.*, vol. 8, pp. 226–228, 1995.

[59] F. Collino and P. Monk, "The perfectly matched absorbing layers in curvilinear coordinates," *SIAM J. Sci. Stat. Comput.*, vol. 119, pp. 2061–2090, 1998. doi:10.1137/S1064827596301406

[60] F. L. Teixeira and W. C. Chew, "Analytical derivation of a conformal perfectly matched absorber for electromagnetic waves," *IEEE Microw. Opt. Technol. Lett.*, vol. 17, pp. 231–236, 1998. doi:10.1002/(SICI)1098-2760(199803)17:4<231::AID-MOP3>3.0.CO;2-J

[61] T. G. Jurgens, J. G. Blackschak, and G. W. Saewert, "Bodies of revolution," in *Computational Electrodynamics. The Finite-Difference Time-Domain Method*, A. Taflove and S. C. Hagness, Eds. Boston, MA: Artech House, 2000, chapter 8.

[62] A. R. Roberts and J. Joubert, "PML absorbing boundary for higher-order schemes," *Electron. Lett.*, vol. 33, pp. 32–34, 1997.

[63] K. Sankaran, C. Fumeaux, and R. Vahldieck, "Cell-centered finite-volume-based perfectly matched layer for time-domain Maxwell system," *IEEE Trans. Microw. Theory Tech.*, vol. 54, pp. 1269–1276, 2006.

[64] C. M. Rappaport, "Interpreting and improving the PML absorbing boundary condition using anisotropic lossy mapping of space," *IEEE Trans. Magn.*, vol. 32, pp. 968–974, 1996.

[65] G. Liu and S. D. Gedney, "Perfectly matched layer media for an unconditionally stable three-dimensional ADI-FDTD method," *IEEE Microw. Guid. Wave Lett.*, vol. 10, pp. 261–263, 2000.

[66] S. D. Gedney, G. Liu, J. A. Roden, and Z. Aiming, "Perfectly matched layer media with CFS for an unconditionally stable ADI-FDTD method," *IEEE Trans. Antennas Propagat.*, vol. 49, pp. 1554–1559, 2001.

[67]   C. Eswarappa and W. J. R. Hoefer, "Implementation of Berenger absorbing boundary conditions in TLM interfacing FDTD perfectly matched layers," *Electron. Lett.*, vol. 31, pp. 1264–1266, 1995.

[68]   N. Pena and M. Ney, "A new TLM nodes for Berenger's perfectly matched layer," *IEEE Microw. Guid. Wave Lett.*, vol. 6, pp. 410–412, 1996.

[69]   J.-L. Dubard and D. Pompei, "Simulation of Berenger's perfectly matched layer with a modified TLM node," *IEE Proc. Microw. Antennas Propagat.*, vol. 144, pp. 205–207, 1997.

[70]   T. Rylander and J.-M. Jin, "Perfectly matched layer for the time domain finite element method," *J. Comput. Phys.*, vol. 200, pp. 238–250, 2004.

[71]   F. Collino, "Perfectly matched absorbing layers for the paraxial equation," *J. Comput. Phys.*, vol. 131, pp. 164–180, 1997.

[72]   M. Levy, *Parabolic Equation Method for Electromagnetic Wave Propagation*. London: The IEE, 2000.

[73]   W. P. Huang, C. L. Xa, W. Li, and K. Yokoyama, "The perfectly matched layer (PML) boundary condition for the beam propagation method," *IEEE Photonics Technol. Lett.*, vol. 8, pp. 649–651, 1996.

[74]   J. Xu, J.-G. Ma, and Z. Chen, "Numerical validation of a nonlinear PML scheme for absorption of nonlinear electromagnetic waves," *IEEE Trans. Microw. Theory Tech.*, vol. 46, pp. 1752–1758, 1998.

[75]   Z. Chen, J. Xu, and J.-G. Ma, "Validation of a nonlinear PML scheme," *IEEE Microw. Guid. Wave Lett.*, vol. 9, pp. 93–95, 1999.

[76]   M. F. Pasik, D. B. Seidel, and R. W. Lemke, "A modified perfectly matched layer implementation for use in electromagnetic PIC codes," *J. Comput. Phys.*, vol. 148, pp. 125–132, 1999.

[77]   Q. Qi and T. L. Geers, "Evaluation of the perfectly matched layer for computational acoustics," *J. Comput. Phys.*, vol. 139, pp. 166–183, 1998.

[78]   X. Yuan, D. Borup, J. W. Wiskin, M. Berggren, R. Eidens, and S. A. Johnson, "Formulation and validation of Berenger's PML absorbing boundary for the FDTD simulation of acoustic scattering," *IEEE Trans. Ultrason. Ferroelectr. Freq. Control*, vol. 44, pp. 816–822, 1997.

[79]   W. C. Chew and Q. H. Liu, "Perfectly matched layer for elastodynamics: A new absorbing boundary condition," *J. Comput. Acoust.*, vol. 4, pp. 341–359, 1996.

[80]   F. Q. Hu, "On absorbing boundary conditions for linearized Euler equations by a perfectly matched layer," *J. Comput. Phys.*, vol. 129, pp. 201–219, 1996.

[81]   F. Q. Hu, "A perfectly matched layer absorbing boundary condition for linearized Euler equations with a non-uniform mean flow," *J. Comput. Phys.*, vol. 208, pp. 469–492, 2005.

Author Biography

Jean-Pierre Bérenger has been with the Centre d'Analyse de Défense (formerly Laboratoire Central de l'Armement), Arcueil, France, since 1975. He received a Master in Physics from the Joseph Fourier University, Grenoble, France, in 1973, and a Master in Optical Engineering from the Institut d'Optique Graduate School (formerly Ecole Supérieure d'Optique), Paris, France, in 1975.

From 1975 to 1984 he was engaged in applied research in the field of the electromagnetic effects of nuclear bursts. During this period he was the author of the DIFRAC computer code, the first FDTD code developed in France for the calculation of the coupling of the nuclear electromagnetic pulse with objects. During years 1984 to 1988 he was involved in the development of simulation software related to ballistic missiles. From 1989 to 1998 he held a position as expert on the electromagnetic effects of nuclear disturbances. He is currently a manager of prospective studies in the field of command, control, and communications.

From 1984 to now, Jean-Pierre Bérenger has stayed active in numerical electromagnetics, in such topics as the FDTD method, absorbing boundary conditions, and low frequency propagation. Most of his works published in the scientific litterature are on the PML absorbing boundary condition and the VLF-LF propagation. In the past fifteen years, he has been an advisor to several laboratories or universities, about the FDTD method and the boundary conditions. He has been also a lecturer on FDTD method in continuing education. He is a senior member of the IEEE, a member of URSI, and a member of the Electromagnetics Academy.